Memory Nanomaterials: Growth, Characterization and Device Fabrication

Memory Nanomaterials: Growth, Characterization and Device Fabrication

Guest Editors

Chao Zhao
Guilei Wang
Huihui Li

MDPI

Basel • Beijing • Wuhan • Barcelona • Belgrade • Novi Sad • Cluj • Manchester

Guest Editors
Chao Zhao
Beijing Superstring Academy of Memory Technology
Beijing
China

Guilei Wang
Beijing Superstring Academy of Memory Technology
Beijing
China

Huihui Li
Beijing Superstring Academy of Memory Technology
Beijing
China

Editorial Office
MDPI AG
Grosspeteranlage 5
4052 Basel, Switzerland

This is a reprint of the Special Issue, published open access by the journal *Nanomaterials* (ISSN 2079-4991), freely accessible at: https://www.mdpi.com/journal/nanomaterials/special_issues/Nano_Memory_Materials.

For citation purposes, cite each article independently as indicated on the article page online and as indicated below:

Lastname, A.A.; Lastname, B.B. Article Title. *Journal Name* **Year**, *Volume Number*, Page Range.

ISBN 978-3-7258-4285-8 (Hbk)
ISBN 978-3-7258-4286-5 (PDF)
https://doi.org/10.3390/books978-3-7258-4286-5

Contents

About the Editors

Chao Zhao

Chao Zhao received PhD degree from Katholieke Universiteit Leuven of Belgium in Material Science and Engineering in 1999. From 2000 to 2010, he worked in IMEC as a senior scientist, and from 2010 to 2021, he served as a full professor in University of Chinese Academy of Sciences and IMECAS. Since 2021, he has been working in Beijing Superstring Academy of Memory Technology as Senior VP. He has also been a part-time professor at Beihang University since 2013.

His research focuses on CMOS and Memory processing technology. He has authored or coauthored over 408 scientific papers, 4 book chapters, and 2 books ("CMOS Past, Present and Future" by Elsevier in 2018 and "Semiconductor Process and Integrated Circuit Manufacturing Technology" by Science Press in 2023) and has been guest editor of two Special Issues, "Advanced Microelectronic and Optoelectronic Materials" (JMSE, 2020) and "Memory Nanomaterials: Growth, Characterization and Device Fabrication (MDPI, 2023)".

He is the fellow of European Academy of Sciences (EurASc), and the Member of the Council and Deputy Secretary General of Chinese Materials Research Society (C-MRS).

Guilei Wang

Guilei Wang received his PhD in 2016 from the University of Chinese Academy of Sciences. He has been worked as a professor at the Institute of Microelectronics of Chinese Academy of Sciences (IMECAS) until 2021. In October 2021, he joined the Beijing Superstring Academy of Memory Technology as a full professor.

His research focuses on new materials, devices, and process integration for the IC industry. So far, more than 160 research papers have been published in international journals. He has completed more than 300 patent applications. He has published 1 book (Investigation on SiGe Selective Epitaxy for Source and Drain Engineering in 22 nm CMOS Technology Node and Beyond), and 1 SiGe epitaxy chapter.

He served as a member of the Technical Program Committee (TPC) of the 2018 European Materials Research Society (E-MRS) Spring Conference. He has been guest editor of three Special Issues. He has won the "E-MRS Young Scientist", "Springer Excellent Doctorate Thesis", and "Excellent Member of the Youth Innovation Promotion Association of the Chinese Academy of Sciences" awards.

Huihui Li

Huihui Li holds a bachelor's degree in theoretical physics from Lanzhou University and a PhD in materials science and engineering from the National University of Singapore. He has previously worked at CAS, CETHIK Group, and Beijing Superstring Academy of Memory Technology, conducting materials and process research related to semiconductor memory technologies such as MRAM, FeRAM, and DRAM. He is currently with Changxin Memory Technologies, working on advanced materials and devices for DRAM.

nanomaterials

Article

Resistive Switching Memory Cell Property Improvement by Al/$SrZrTiO_3$/Al/$SrZrTiO_3$/ITO with Embedded Al Layer

Ke-Jing Lee [1,2], Wei-Shao Lin [1], Li-Wen Wang [1], Hsin-Ni Lin [3] and Yeong-Her Wang [1,2,*]

1 Institute of Microelectronics, Department of Electrical Engineering, National Cheng-Kung University, Tainan 701, Taiwan
2 Program on Semiconductor Process Technology, Academy of Innovative Semiconductor and Sustainable Manufacturing, National Cheng-Kung University, Tainan 701, Taiwan
3 Department of Physics, National Sun Yat-sen University, Kaohsiung 804, Taiwan
* Correspondence: yhw@ee.ncku.edu.tw; Tel.: +886-6-275-7575-62352; Fax: +886-6-2080598

Abstract: The $SrZrTiO_3$ (SZT) thin film prepared by sol-gel process for the insulator of resistive random-access memory (RRAM) is presented. Al was embedded in the SZT thin film to enhance the switching characteristics. Compared with the pure SZT thin-film RRAM, the RRAM with the embedded Al in SZT thin film demonstrated outstanding device parameter improvements, such as a resistance ratio higher than 10^7, lower operation voltage ($V_{SET} = -0.8$ V and $V_{RESET} = 2.05$ V), uniform film, and device stability of more than 10^5 s. The physical properties of the SZT thin film and the embedded-Al SZT thin-film RRAM devices were probed.

Keywords: resistive random-access memory (RRAM); sol-gel; strontium zirconate titanium

1. Introduction

Among various candidates for nonvolatile memories, resistive random-access memory (RRAM) attracts considerable attention owing to its advantages, such as simple metal–insulator–metal (MIM) structure, fast operation speed, low operation voltage, and low fabrication temperature. Numerous materials, such as transition metal oxides (HfO_2, TiO_2, and NiO), organic material (composites containing nanoparticles and polyimide), or perovskite oxides ($SrZrO_3$ and $SrTiO_3$), have been investigated as potential materials for RRAMs. In this study, we focused on the $SrTiO_3$ material because of its high dielectric constant, low dielectric loss, tenability, high breakdown strength, low leakage current, and great film quality [1]. Numerous attempts have been exerted to further improve the properties of $SrTiO_3$-based ceramics. Doping is considered an effective approach for altering their properties [2]. Reproducible resistive switching behaviors have been observed in doped perovskite oxide films [3,4], such as Mo-doped $SrTiO_3$ [5], Mg-doped $SrTiO_3$ [6], Nb-doped $SrTiO_3$ films [7], and Ni-doped $SrTiO_3$ films [3]. Several studies have indicated that Zr^{4+} in $SrTiO_3$ can stabilize the charge of Ti^{4+} and suppress oxygen dissociation [1,2,8]. In addition, our previous study showed that the addition of Zr can effectively improve the surface morphology of insulators [9]. Moreover, the sol-gel process has advantages, such as low fabrication temperature, low cost, and easy adjustment of proportions. This process can be applied in numerous devices [10]. Based on the above advantages and research results, in this work, we utilized Zr in the doping process to fabricate $SrZrTiO_3$ (SZT) using the sol-gel process for RRAM insulators.

Various inserted metal layers in the insulator layer, including Cu, Pt, and Ti, improve the resistive switching properties of RRAMs [11–14]. However, a relatively limited number of studies have been performed regarding the mechanism of embedded metal-based RRAM devices. In this study, the effects of Al-included SZT on the resistive switching properties of SZT thin film for RRAM applications were also investigated. For comparison, a memory device with a single-SZT active layer was also fabricated and characterized.

2. Materials and Methods

The solutions used in this study were prepared in reference to the previous literature, and the additives and ratios were varied based on experimental requirements [8,15,16]. The 0.5 M SZT solution was prepared in three steps (Figure 1a). First, strontium acetate (537 mg) was dissolved in glacial acetic acid (4 mL) by stirring and then heated on a 100 °C hot plate until complete dissolution (A1 solution). Titanium isopropoxide (760 μL) and zirconium n-propoxide (770 μL) were mixed with acetylacetone (470 μL) and dissolved in 2-methoxyethanol (7.74 mL) by stirring (A2 solution). Finally, A1 was slowly dropped into A2 until a completely transparent solution was obtained (mole proportion of Sr:Zr:Ti was approximately 1:1:1). After chemically cleaning the substrate, the prepared 0.5 M SZT solution was spun coated on an indium tin oxide (ITO)/glass substrate. Each layer was baked at 100 °C for 15 min to remove volatile materials. A 90 nm-thick Al film with an area of 3 mm^2 was deposited with a shadow mask using DC magnetron sputtering as the top electrode (TE) of the MIM structure, whereas ITO served as the bottom electrode (BE). Figure 1b shows the schematic configuration of the Al/SZT/ITO structures.

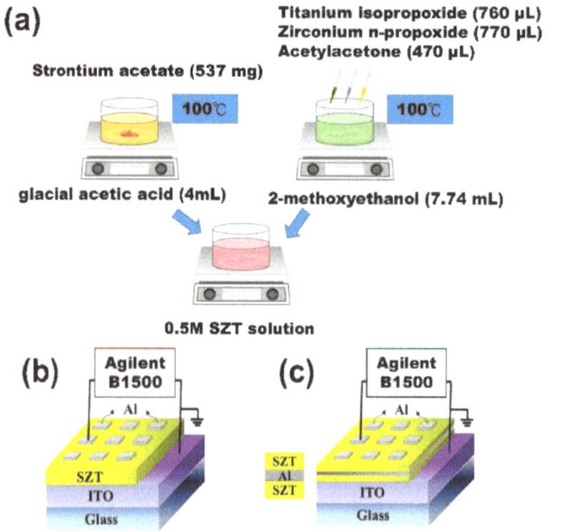

Figure 1. (**a**) Preparation of 0.5 M SZT solution. (**b**) Schematic of SZT-based RRAM devices. (**c**) Schematic of Al embedded in SZT thin-film RRAM devices.

The SZT/Al/SZT tri-layer structure was deposited on the ITO/glass substrate. Both SZT films were prepared through the same process above, and Al was embedded by DC magnetron sputtering. Finally, the same processes were applied for Al TEs. For comparison, we fabricated four different thicknesses of embedded Al layer (5, 18, 25, and 33 nm). Figure 1c displays the schematic configuration of embedded Al in SZT thin-film RRAM devices. Transmission electron microscopy (TEM) analysis was carried out using a JEM-2100F electron microscope. The X-ray diffraction (XRD) spectra were developed via BRUKER, D8 DISCOVER SSS Multi-Function High Power XRD using the copper Kα line with λ = 0.154060 nm. The roughness value and surface morphology of the films were calculated using atomic force microscopy (AFM, Dimension ICON with NanoScope V controller, Bruker, Billerica, MA, USA). X-ray photoelectron spectroscopy (XPS) was performed using a PHI 5000 VersaProbe. Electrical measurements were performed using an Agilent B1500 semiconductor parameter analyzer.

3. Results

Figure 2a presents the TEM image of the cross-sectional SZT/Al/SZT tri-layered thin films sandwiched between ITO BE and Al TE. The estimated thicknesses of the top SZT layer, Al interlayer, and bottom SZT layer were 54, 18, and 67 nm, respectively. The thin films were uniform, and the interfaces were clear. Figure 2b depicts the XRD analysis of the SZT thin film. As shown in the analysis results, no evident peak was observed in the XRD spectrum. The SZT thin film prepared in this experiment had an amorphous phase. Compared with the polycrystalline phase, the amorphous phase is preferred for dielectric layer materials because the former may lead to a high-grain boundary leakage current and a rough film surface [17]. The AFM images in Figure 2c,d reveal the surface morphology of the SZT and SZT/Al/SZT tri-layered thin films. The root-mean-square roughness (R_{rms}) values of SZT and the SZT/Al/SZT thin film were approximately 2.0 and 0.8 nm, respectively. The roughness of the thin films decreased due to the insertion of Al layer.

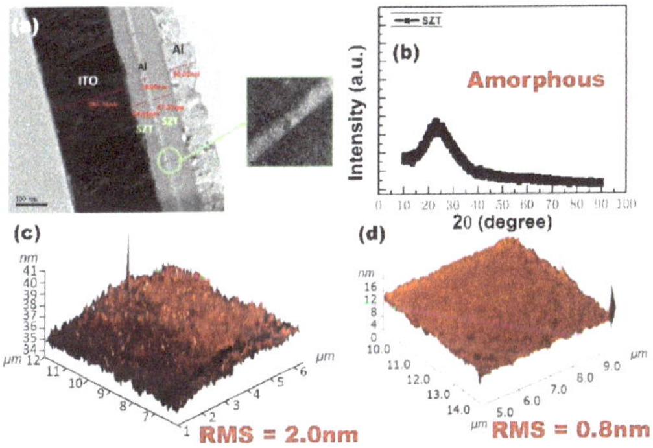

Figure 2. (**a**) TEM image of cross-sectional SZT/Al/SZT tri-layered thin films sandwiched between ITO BE and Al TE. (**b**) XRD spectrum of SZT thin film. AFM image of (**c**) SZT and (**d**) SZT/Al/SZT tri-layered thin films.

The XPS analysis was used to obtain the chemical composition of SZT thin film, as shown in Figure 3a. The atomic percentage of Sr, Zr, Ti, and O were obtained as 13%, 11%, 16%, and 60%, respectively. By fitting the peaks of Sr 3d signal can be decomposed into two peaks of Sr $3d_{5/2}$ and Sr $3d_{3/2}$, which are centered at 132.7 and 134.1 eV, respectively. The peak at 132.7 eV can be attributed to $SrTiO_3$, and the peak at 134.1 eV can be attributed to $SrCO_3$. By fitting the peaks of Zr3d signal can be decomposed into two peaks of Zr3d5/2 and Zr $3d_{3/2}$, which are centered at 182 and 184.4 eV, respectively. This result confirms that most of Zr atoms are incorporated at Ti lattice site instead of forming ZrO_2 [18,19]. The binding energy around 182 and 184.4 eV are assigned for $ZrTiO_4$. The peaks for the O1s signal may be consistently fitted by three different near–Gaussian sub-peaks centered at 529.2, 530.9, and 532.1 eV. The binding energy of lattice oxygen is 529.2 eV, which is attributed to the O_2 ions bonded with Sr, Zr, and Ti ions. The peaks located at 530.9 eV were associated with non-lattice oxygen ions, such as oxygen vacancy, and the peak at 532.1 eV corresponds to surface adsorption oxygen in the $SrTiO_3$ catalyst [3,9,20]. Peaks fitting analyses of the Ti 2p signal can be decomposed into two peaks of Ti $2p_{3/2}$ and Ti $2p_{1/2}$, with two components of the binding energies at 458.8 eV and 464.3 eV, which are attributed to $SrTiO_3$ and $ZrTiO_4$. To investigate the possible inter-diffusion in the SZT/Al/SZT tri-layer, we determined the XPS depth profile of the SZT/Al/SZT tri-layer thin film (Figure 3b). The Al concentration increased, and Sr, Zr, and Ti concentrations correspondingly decreased with depth after 36 s Ar^+ sputtering. This result demonstrated the significant interfacial diffusion between SZT/Al/SZT, which is in good agreement with the TEM cross-sectional image. In spite of this finding, the memory units of the SZT/Al/SZT tri-layer structure on ITO-coated glass have been fabricated by sputtering [21,22].

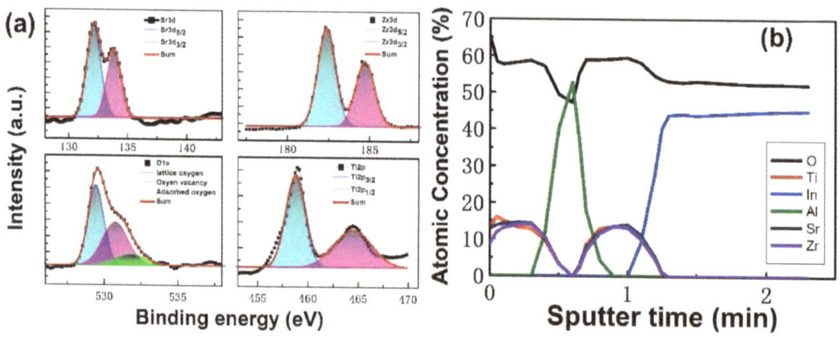

Figure 3. (**a**) High-resolution XPS spectra of Sr3d, Zr3d, Ti2p, and O1s of the SZT thin film. (**b**) Atomic concentrations of six elements in the SZT/Al/SZT tri-layered thin film with XPS depth profiling.

Figure 4a depicts the bipolar and reproducible resistive switching behaviors of the Al/SZT/ITO structures. During the negative sweep from 0 to −4 V, the current increased sharply, a phenomenon called the "set" process, switching from a high resistance state (HRS) to a low resistance state (LRS). Sequentially, during the positive sweep from 0 to 4 V, an abrupt decrease in current, that is, the "reset" process, was observed, and it switched the resistance state from LRS to HRS with 10 mA compliance current. The V_{SET}/V_{RESET} of Al/SZT/ITO was −1.28 V/+2.08 V. The ON/OFF ratio of the devices was around 10^3. Figure 4b shows the Al/SZT/SZT/ITO RRAM devices. The set voltage (V_{SET}) was −2.88 V, and the reset voltage (V_{RESET}) was located at +2.48 V. The ON/OFF ratio of the devices was around 10. Figure 4c–f shows the discussed resistive switching I–V characteristic of different thicknesses (5, 18, 25, and 33 nm) of the embedded Al. Therefore, the observed I–V characteristic of the 18 nm embedded Al displayed an especially outstanding performance. However, the memory margins of embedded Al layer with different thicknesses were insufficient for memory application. The V_{SET}/V_{RESET} of Al/SZT/Al (18 nm)/SZT/ITO was −0.32/+1.28 V (Figure 4d). The ON/OFF ratio of the devices was around 10^7. No significant switching behavior was observed in the embedded structure with different thicknesses (5, 25, and 33 nm) of Al (Figure 4c,e,f).

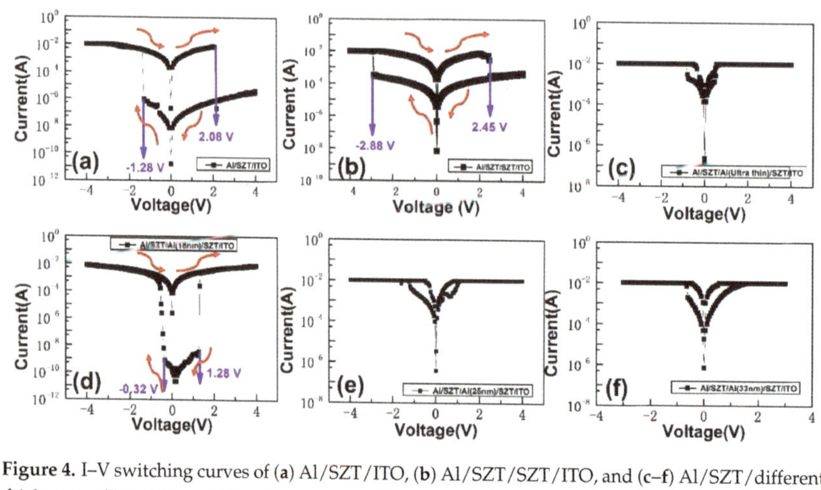

Figure 4. I–V switching curves of (**a**) Al/SZT/ITO, (**b**) Al/SZT/SZT/ITO, and (**c**–**f**) Al/SZT/different thicknesses (5, 18, 25, and 33 nm) of embedded metal –Al /SZT/ITO device RRAMs.

To probe the mechanisms of resistive switching characteristics of SZT-based RRAM, we performed the curve fittings of conduction mechanisms for HRS and LRS in the

Al/SZT/ITO and Al/SZT/Al (18 nm)/SZT/ITO devices, and the I–V characteristic was plotted in log-log scale (Figure 5a,b). At low applied voltage, the slope of HRS was very close to linear. As the voltage increased, the current of HRS followed a voltage-square dependence. With the continuous increase in bias voltage, the current of HRS increased rapidly, corresponding to the steep increase in the current region. After the abrupt increase in its region, the current showed voltage-square dependence again. The fitting results of HRS illustrated that the current showed typical space-charge-limited conduction, which consists of the ohmic region ($I \propto V$), trap-filled limit current ($I \propto V^2$), and Child's law region ($I \propto V^2$) [23,24].

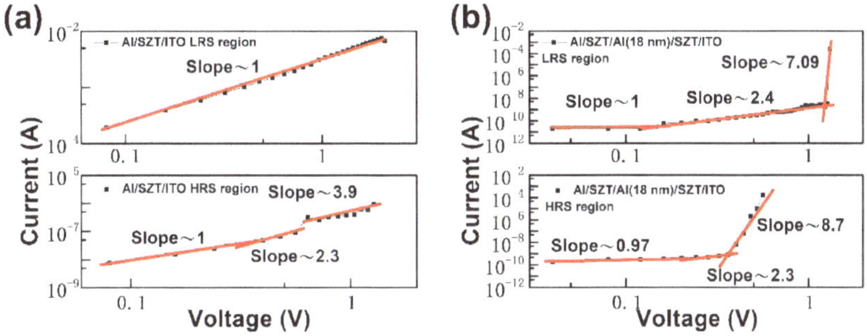

Figure 5. Double logarithmic plot and linear fitting of the switching I–V curve of (**a**) Al/SZT/ITO and (**b**) Al/SZT/Al (18 nm)/SZT/ITO device RRAMs.

However, the conduction behaviors of the LRS showed distinct features for Al/SZT/ITO and Al/SZT/Al (18 nm)/SZT/ITO devices. By contrast, the curve fitting of LRS of Al/SZT/ITO device showed the ohmic conduction behavior, coinciding with a conducting filamentary model [25,26]. The I-V curve of the LRS of Al/SZT/Al (18 nm)/SZT/ITO device consisted of the ohmic region ($I \propto V$) and trap-filled limit current ($I \propto V^2$) and Child's law region ($I \propto V^2$), which indicates that resistance switching in Al/SZT/Al (18 nm)/SZT/ITO device was mediated by a carrier trapping/de-trapping process [27]. They can remain below V_{RESET} in LRS Part 3 but reset did not occur until V_{RESET} was reached again in HRS Part 4 because of the lag induced by the relaxation of trap-filled states (Figure 4d). The simple embedded Al layer not only remarkably improved the device parameters but essentially altered the switching mechanism.

The trap depth (ψ_t) of ~0.91 eV for the SZT memory device at the HRS can be extrapolated from an intercept of ln(J/E) as the function of temperature (Figure 6) [24,28]. $\psi_{t,A}$ and $\psi_{t,B}$ are the trap depth of the Al/SZT/ITO and Al/SZT/Al (18 nm)/SZT/ITO structure, respectively. The equivalent trap depth can be reduced to ~0.44 eV after the insertion of Al layer. The possible carrier transporting according to the carriers hops along the trapping states. Under the same barrier height condition, the reduction of trap depth enhances opportunities for electron hopping, thereby significantly reducing the HRS current and leading to a high ON/OFF ratio of 10^7. From the above results, initially, native defects of oxygen vacancies scatter in the SZT film. Injected electrons are trapped in the defects and affect the current conduction. Thus, the conduction mechanism in HRS is dominated by SCLC.

Figure 7a shows the Al/SZT/Al (18 nm)/SZT/ITO structure band diagram under zero bias. The valleys on the upper side of the SZT insulator represent the traps inside the SZT insulator. The left SZT insulator layer was leaned as the result of the different work functions between ITO and Al (ITO: 4.5 eV and Al: 4.2 eV). The Al layer embedded in the SZT thin film acted as a trap for electrons because the electron affinity of SZT was higher than the work function of Al. When the devices were under low negative bias, which was equivalent to the Al TE, a negative bias was applied, and the ITO bottom was under applied positive bias (Figure 7b). In the low applied-bias negative region, the transport in RRAM devices was caused by thermally generated carriers [11]. When the carriers crossed over the first SZT insulator from the Al TE, they were trapped inside the embedded Al layer. When the applied negative voltage increased, the traps in the SZT insulator were almost filled, the carriers began to transport in the SZT insulator, and the concentration of carriers inside the SZT insulator increased. Meanwhile, the carriers trapped in the embedded Al layer also joined the transport. In this region, the current steeply increased, and the RRAM switched from HRS to LRS. By contrast, the switch of devices from LRS to HRS was also discussed. First, when the applied positive was small, the carriers were gradually released by the trap in the SZT insulator, that is, the detrapping process (Figure 7c). When the positive arrives at the V_{RESET}, the devices originally should switch from LRS to HRS. However, given the carriers trapped inside the embedded Al layer, the carrier concentration in the SZT insulator remained high, and the devices were maintained in the LRS state. During the applied positive sweeps from +4 V to the V_{RESET} (about +1.28 V), the RRAM devices reached the detrapping situation. The carriers could not fill the trap, and the carrier concentration in the SZT insulator decreased, causing the devices to switch from the LRS to HRS. As a result, the embedded Al layer structures in RRAM devices caused a decrease in the HRS current and resulted in a high on/off ratio. Further, the embedded Al layer structures in RRAM devices caused a lag in the transport process, and we can discover the phenomenon in the I–V characteristic.

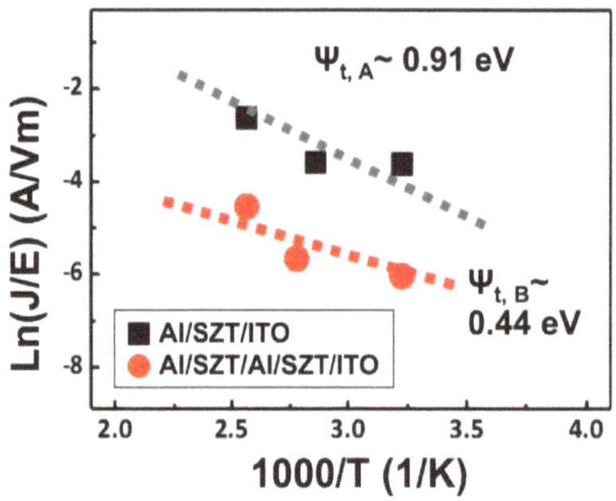

Figure 6. ln(J/E) as the function 1/T of the Al/SZT/ITO and Al/SZT/Al (18 nm)/SZT/ITO structure.

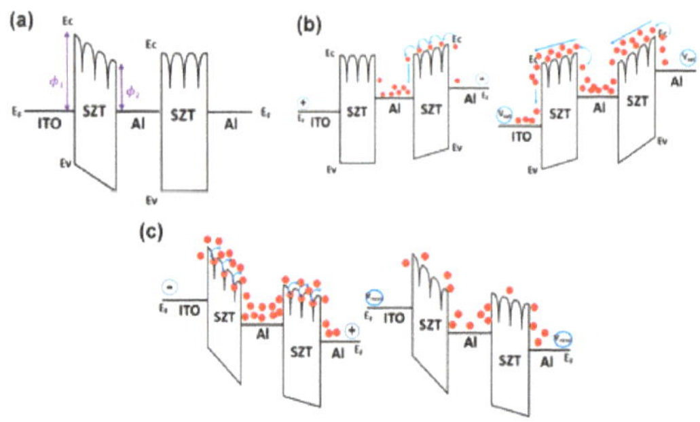

Figure 7. Possible resistive switching mechanism of Al/SZT/Al (18 nm)/SZT/ITO devices. (**a**) Thermal equilibrium, (**b**) negative bias, and (**c**) positive bias.

Figure 8 shows the reliability issues of Al/SZT/ITO and Al/SZT/Al (18 nm)/SZT/ITO structures. The distribution of current and operation voltage was investigated. The V_{SET} and V_{RESET} distribution were plotted in a histogram to demonstrate the reliability of the

Al/SZT/ITO and Al/SZT/Al (18 nm)/SZT/ITO RRAM (Figure 8a,b, respectively). In mathematics, μ is the mean, and σ is the standard deviation [29]. As presented in Figure 8a, the V_{SET} and V_{RESET} were distributed widely. The average and standard deviation values of the V_{SET} were −1.46 and 0.36, respectively. The average and standard deviation values of V_{RESET} were 2.46 and 0.36 V, respectively. As shown in Figure 8b, when the devices switched between ON and OFF states, V_{SET} was distributed in the range of −0.2 to −0.68 V, whereas V_{RESET} was distributed in the range of 0.92 to 1.52 V. The average and standard deviation values of V_{SET} were −0.44 and 0.18 V, respectively. The average and standard deviation values of V_{RESET} were 1.22 and 0.22 V, respectively. The reduction and more stable distribution of V_{SET} and V_{RESET} may be related to the addition of an 18 nm embedded Al layer [30].

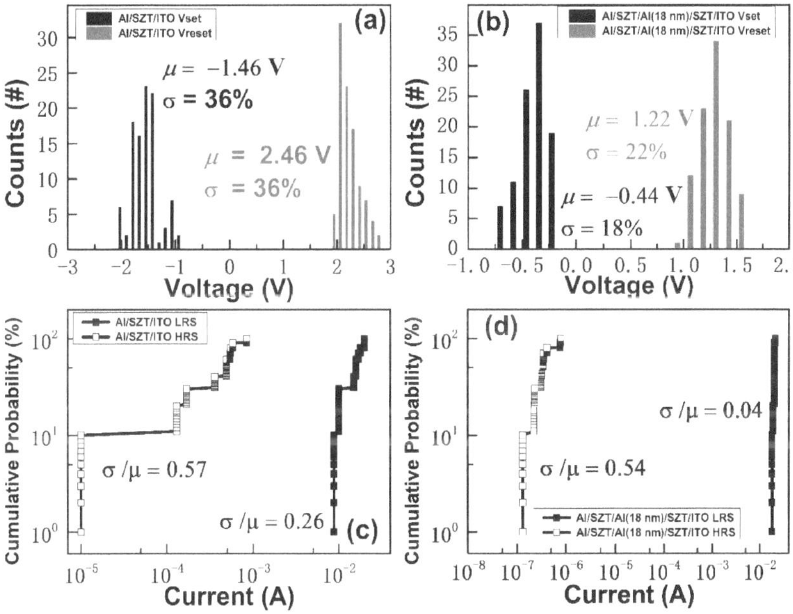

Figure 8. Statistical and cumulative probability distribution of V_{SET} and V_{RESET} measured from (**a**) Al/SZT/ITO and (**b**) Al/SZT/Al (18 nm)/SZT/ITO devices unit during 100 time tests. The distribution of HRS/LRS resistance of (**c**) Al/SZT/ITO and (**b**) Al/SZT/Al (18 nm)/SZT/ITO devices unit for 100 time tests.

Figure 8c,d illustrates the cumulative probabilities of HRS/LRS resistance for the Al/SZT/ITO and Al/SZT/Al (18 nm)/SZT/ITO device. Comparing the 18 nm embedded Al RRAM with the SZT-based RRAM, the coefficient declined from 0.26 to 0.04 at LRS and decreased by 0.03% at HRS, which reveals the 18 nm embedded Al RRAM depicted a large memory window and uniformity compared with SZT-based RRAM.

Figure 9a plots the scaling trend of the LRS resistance versus the cell area of the Al/SZT/ITO and Al/SZT/Al (18 nm)/SZT/ITO structures. The LRS resistance is mainly a filamentary conduction current, and thus, it only had a slight dependence on the cell area. These results are similar to those for other metal oxides [31]. Figure 9b presents the retention capabilities of Al/SZT/ITO and Al/SZT/Al (18 nm)/SZT/ITO structures measured at room temperature at a voltage of 0.5 V. The ON/OFF ratio of the Al/SZT/Al (18 nm)/SZT/ITO structure remained higher than 10^6, and that for the Al/SZT/ITO structure was around 10^3. For the Al/SZT/Al (18 nm)/SZT/ITO structure, the current magnitudes did not differ significantly over 10^5 s. The smooth roughness may facilitate stable resistive switching [3], and the stability of the switching cycle and retention capability improved compared with those of the SZT memory device.

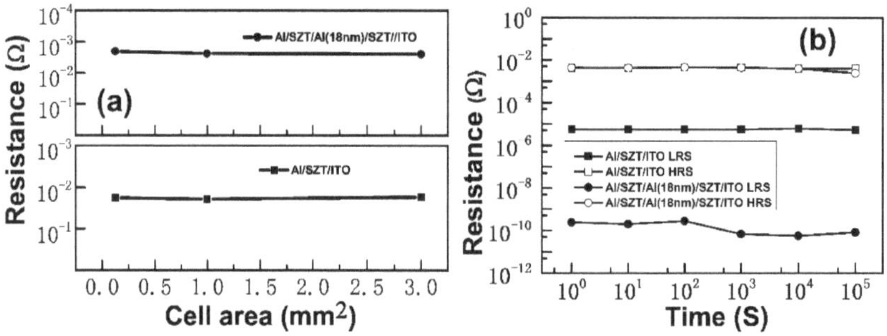

Figure 9. (**a**) LRS current with cell sizes ranging from 0.125 mm^2 to 3 mm^2 and (**b**) retention capability of Al/SZT/ITO and Al/SZT/Al (18 nm)/SZT/ITO structures.

From Table 1, the I–V switching characteristics of RRAM devices were promoted by inserting the Al embedded layer. Compared with the pure SZT-based RRAM devices and reported data [11–14], the resistance ratio increased from 10^3 to 10^7, which is an ultrahigh memory margin, which led to distinguishing the storage information easily. The V_{SET} decreased from −1.28 to −0.32 V, and the V_{RESET} from 2.08 to 1.28 V. These results indicate that the operation voltage and current can be decreased to lower power consumption.

Table 1. Performance comparison of the SZT-based RRAM devices with reported data.

Insulator Material	Insert Metal	Fabrication of the Insulator	Fabrication of the Insert Metal	Resistance Ratio	Vest (V)	Vreset (V)	Current of HRS (A)	Ref
ZnO	Cu	Sputter	Sputter	10^4	0.9	−0.6	10^{-4}	[11]
TiOx	Pt	Thermal oxidation	–	10^5	3.7	−0.9	10^{-8}	[12]
HfO_2	Ti	Sputter	Sputter	50	0.7	−1.3	10^{-3}	[13]
ZrO_2	Ti	Electron-beam evaporation	Implant	10^4	1.3	−0.66	10^{-9}	[14]
SZT	–	Sol-gel	Sputter	10^3	−1.28	2.08	10^{-8}	This work
SZT	Al	Sol-gel	Sputter	10^7	−0.32	1.28	10^{-10}	This work

4. Conclusions

In conclusion, resistive switching behaviors in sol-gel SZT thin films have been probed, and the bipolar resistive switching characteristics of Al/SZT/ITO devices can utilize Al as an embedded layer in the SZT thin film for enhancing properties. Significant improvements in the ON/OFF ratios from 10^3 to over 10^7 were observed upon application to the Al/SZT/Al (18 nm)/SZT/ITO structure without the forming processes. A high interface trap density with a large amount of trapped electrons will help in building a favorable electric field to attract oxygen vacancies, and thus, the lower V_{SET} and V_{RESET} compared with those of the SZT memory device were obtained. The reduced trapping depths from 0.91 eV to 0.44 eV were found after the insertion of the Al layer, resulting in a low HRS current or high ON/OFF ratio. Therefore, the free-forming Al/SZT/Al (18 nm)/SZT/ITO structure with a high ON/OFF ratio of over 10^7, excellent voltage distribution, and good retention of over 10^5 s can be achieved.

Author Contributions: K.-J.L. was responsible for the device preparation and characterization, data analysis, modeling discussion, and writing and editing of the manuscript. W.-S.L. was responsible for data analysis. L.-W.W. and H.-N.L. were responsible for the characterization and modeling discussion. Y.-H.W. monitored the progress and paper editing. All authors have read and agreed to the published version of the manuscript.

Funding: This work was supported in part by the Ministry of Science and Technology of Taiwan under grant MOST 108-2221-E-006-040-MY3. The authors would like to thank the Core Facility Center, Micro/Nano Technology Group, National Cheng-Kung University, Tainan, Taiwan, and Taiwan Semiconductor Research Institute, National Applied Research Laboratories, Taiwan for the support of high frequency and flicker noise measurements.

Institutional Review Board Statement: Not applicable.

Informed Consent Statement: Not applicable.

Data Availability Statement: The data presented in this study are available upon request from the corresponding author.

Acknowledgments: We appreciate the Core Facility Center, Micro/Nano Technology Group, National Cheng-Kung University, Tainan, Taiwan (device fabrication and characterization) and Taiwan Semiconductor Research Institute, National Applied Research Laboratories, Taiwan for their support including high frequency measurements.

Conflicts of Interest: The authors declare no conflict of interest.

References

1. Muhamad, N.F.; Osman, R.A.M.; Idris, M.S.; Yasin, M.N.M. Physical and electrical properties of $SrTiO_3$ and $SrZrO_3$. *EPJ Web Conf.* **2017**, *162*, 01052. [CrossRef]
2. Muhamad, N.F.; Osman, R.A.M.; Idris, M.S.; Jamlos, F.; Hambali, N.A.M.A. Microwave and Electrical Properties of Zr-Doped $SrTiO_3$ for Dielectric Resonator Antenna Application. *Solid State Phenom.* **2018**, *280*, 142–148. [CrossRef]
3. Lee, K.-J.; Chang, Y.-C.; Lee, C.-J.; Wang, L.-W.; Chou, D.-W.; Chiang, T.-K.; Wang, Y.-H. Effects of Ni in Strontium Titanate Nickelate Thin Films for Flexible Nonvolatile Memory Applications. *IEEE Trans. Electron Devices* **2017**, *64*, 2001–2007. [CrossRef]
4. Chen, G.; Zhang, P.; Pan, L.; Qi, L.; Yu, F.; Gao, C. Flexible nonvolatile resistive memory devices based on $SrTiO_3$ nanosheets and polyvinylpyrrolidone composites. *J. Mater. Chem. C* **2017**, *5*, 9799–9805. [CrossRef]
5. Padmini, E.; Ramachandran, K. Electrical properties of Mo-doped $SrTiO_3$ prepared by solid state reaction method. *Mater. Res. Express* **2019**, *6*, 115919. [CrossRef]
6. Zhang, W.; Wang, H.; Xu, J.; Liu, G.; Xie, H.; Yang, L. Effects of Mg Doping Concentration on Resistive Switching Behavior and Properties of $SrTi_{1-y}Mg_yO_3$ Films. *J. Wuhan Univ. Technol.-Mater. Sci. Ed.* **2019**, *34*, 888–892. [CrossRef]
7. Zhang, T.F.; Tang, X.G.; Liu, Q.X.; Jiang, Y.P. Electrode effect regulated resistance switching and selector characteristics in Nb doped $SrTiO_3$ single crystal for potential cross-point memory applications. *J. Alloys Compd.* **2018**, *730*, 516–520. [CrossRef]
8. Bera, J.; Rout, S.K. $SrTiO_3$–$SrZrO_3$ solid solution: Phase formation kinetics and mechanism through solid-oxide reaction. *Mater. Res. Bull.* **2005**, *40*, 1187–1193. [CrossRef]
9. Wei, C.Y.; Huang, W.C.; Yang, C.K.; Chang, Y.Y.; Wang, Y.H. Low-operating-voltage pentacene-based transistors and inverters with solution-processed barium zirconate titanate insulators. *IEEE Electron Device Lett.* **2011**, *32*, 1755–1757. [CrossRef]
10. Lee, K.-J.; Wang, Y.-H. Effect of Alkaline Earth Metal on $AZrO_x$ (A = Mg, Sr, Ba) Memory Application. *Gels* **2022**, *8*, 20. [CrossRef]
11. Yang, Y.C.; Pan, F.; Zeng, F.; Liu, M. Switching mechanism transition induced by annealing treatment in nonvolatile Cu/ZnO/Cu/ZnO/Pt resistive memory: From carrier trapping/detrapping to electrochemical metallization. *J. Appl. Phys.* **2009**, *106*, 123705-1–123705-5. [CrossRef]
12. Chung, Y.L.; Lai, Y.; Chen, Y.C.; Chen, J.S. Schottky barrier mediated single-polarity resistive switching in Pt layer-included TiOx memory device. *ACS Appl. Mater. Interfaces* **2011**, *3*, 1918–1924. [CrossRef] [PubMed]
13. Lee, D.; Sung, Y.; Lee, I.; Kim, J.; Sohn, H.; Ko, D.H. Enhanced bipolar resistive switching of HfO_2 with a Ti interlayer. *Appl. Phys. A* **2011**, *102*, 997–1001. [CrossRef]
14. Liu, Q.; Long, S.; Wang, W.; Zuo, Q.; Zhang, S.; Chen, J.; Liu, M. Improvement of resistive switching properties in ZrO_2-based ReRAM with implanted Ti ions. *IEEE Electron Device Lett.* **2009**, *30*, 1335–1337. [CrossRef]
15. Ruth, E.; McKnight, A.; Kennedy, B.J.; Zhou, Q.; Carpenter, M.A. Elastic anomalies associated with transformation sequences in perovskites: II. The strontium zirconate–titanate Sr (Zr, Ti)O_3 solid solution series. *J. Phys. Condens. Matter* **2008**, *21*, 015902. [CrossRef]
16. Dixit, D.; Agrawal, C.; Mohapatra, Y.N.; Majumder, S.B.; Katiyar, R.S. Studies on the dielectric and relaxor behavior of sol–gel derived barium strontium zirconate titanate thin films. *Mater. Lett.* **2007**, *61*, 3685–3688. [CrossRef]
17. Huang, S.; Lee, K.-J.; Wang, Y.-H. Magnesium Zirconate Titanate Thin Films Used as an NO_2 Sensing Layer for Gas Sensor Applications Developed Using a Sol–Gel Method. *Sensors* **2021**, *21*, 2825. [CrossRef] [PubMed]
18. Avilés, M.A.; Córdoba, J.M.; Sayagués, M.J.; Gotor, F.J. Mechanochemical synthesis of $Ti_{1-x}Zr_xB_2$ and $Ti_{1-x}Hf_xB_2$ solid solutions. *J. Phys. Chem. B* **1998**, *102*, 5094–5098. [CrossRef]
19. Victor, S.B. Krupanidhi, Impact of microstructure on electrical characteristics of laser ablation grown $ZrTiO_4$ thin films on Si substrate. *J. Phys. D. Appl. Phys* **2005**, *38*, 41–50. [CrossRef]

20. Praveen, T.; Rose, T.P.; Saji, K.J. Top electrode dependent resistive switching in M/ZnO/ITO memristors, M = Al, ITO, Cu, and Au. *Microelectron. J.* **2022**, *121*, 105388. [CrossRef]
21. Zhang, W.; Kong, Z.J.; Cao, Y.Z.; Li, D.A.; Wang, G.L.; Zhu, L.; Li, X.; Cao, Y.-Q.; Wu, D. Bipolar Resistive Switching Characteristics of $HfO_2/TiO_2/HfO_2$ Trilayer-Structure RRAM Devices on Pt and TiN-Coated Substrates Fabricated by Atomic Layer Deposition. *Nanoscale Res. Lett.* **2017**, *12*, 393. [CrossRef] [PubMed]
22. Saleem, F.; Simanjuntak, M.; Chandrasekaran, S.; Rajasekaran, S.; Tseng, T.-Y.; Prodromakis, T. Transformation of digital to analog switching in TaO_x-based memristor device for neuromorphic applications. *Appl. Phys. Lett.* **2021**, *118*, 112103. [CrossRef]
23. Lampert, M.A. Simplified Theory of Space-Charge-Limited Currents in an Insulator with Traps. *Phys. Rev. J. Arch.* **1956**, *103*, 1648. [CrossRef]
24. Chiu, F.C. A review on conduction mechanisms in dielectric films. *Adv. Mater. Sci. Eng.* **2014**, *2014*, 1–18. [CrossRef]
25. Wang, Z.S.; Zeng, F.; Yang, J.; Chen, C.; Yang, Y.C.; Pan, F. Reproducible and controllable organic resistive memory based on Al/poly(3,4-ethylene-dioxythiophene):poly(styrenesulfonate)/Al structure. *Appl. Phys. Lett.* **2010**, *97*, 253301. [CrossRef]
26. Yang, Y.C.; Pan, F.; Liu, Q.; Liu, M.; Zeng, F. Fully Room-Temperature-Fabricated Nonvolatile Resistive Memory for Ultrafast and High-Density Memory Application. *Nano Lett.* **2009**, *9*, 1636–1643. [CrossRef]
27. Park, K.H.; Jung, J.H.; Li, F.; Son, D.I.; Kim, T.W. Carrier transport mechanisms of nonvolatile memory devices based on nanocomposites consisting of ZnO nanoparticles with polymethylmethacrylate nanocomposites sandwiched between two C60 layers. *Appl. Phys. Lett.* **2008**, *93*, 132104. [CrossRef]
28. Sedghi, N.; Davey, W.; Mitrovic, I.Z.; Hall, S. Design and focused ion beam fabrication of single crystal diamond nanobeam cavities. *J. Vac. Sci. Technol.* **2011**, *B29*, 01AB10. [CrossRef]
29. Ryu, S.W.; Cho, S.; Park, J.; Kwac, J.; Kim, H.J.; Nishi, Y. Effects of ZrO_2 doping on HfO_2 resistive switching memory characteristics. *Appl. Phys. Lett.* **2014**, *105*, 072102. [CrossRef]
30. Chen, L.; Dai, Y.W.; Sun, Q.Q.; Guo, J.J.; Zhou, P.; Zhang, D.W. Al_2O_3/HfO_2 functional stack films based resistive switching memories with controlled SET and RESET voltages. *Solid State Ion.* **2015**, *273*, 66–69. [CrossRef]
31. Wong, H.S.; Lee, H.Y.; Yu, S.; Chen, Y.S.; Wu, Y.; Chen, S.; Lee, B.; Chen, F.T.; Tsai, M.J. Metal-oxide RRAM. *Proc. IEEE* **2012**, *100*, 1951–1970. [CrossRef]

nanomaterials

MDPI

Article

Comparative Study of Temperature Impact in Spin-Torque Switched Perpendicular and Easy-Cone MTJs

Jingwei Long [1,†], Qi Hu [2,†], Zhengping Yuan [1,†], Yunsen Zhang [2], Yue Xin [1], Jie Ren [1], Bowen Dong [2], Gengfei Li [2], Yumeng Yang [1,3], Huihui Li [2,*] and Zhifeng Zhu [1,3,*]

1 School of Information Science and Technology, Shanghai Tech University, Shanghai 201210, China
2 Beijing Superstring Academy of Memory Technology, Beijing 100176, China
3 Shanghai Engineering Research Center of Energy Efficient and Custom AI IC, Shanghai 201210, China
* Correspondence: huihui.li@bjsamt.org.cn (H.L.); zhuzhf@shanghaitech.edu.cn (Z.Z.)
† These authors contributed equally to this study.

Abstract: The writing performance of the easy-cone magnetic tunnel junction (MTJ) and perpendicularly magnetized MTJ (pMTJ) under various temperatures was investigated based on the macrospin model. When the temperature is changed from 273 K to 373 K, the switching current density of the pMTJ changes by 56%, whereas this value is only 8% in the easy-cone MTJ. Similarly, the temperature-induced variation of the switching delay is more significant in the pMTJ. This indicates that the easy-cone MTJ has a more stable writing performance under temperature variations, resulting in a wider operating temperature range. In addition, these two types of MTJs exhibit opposite temperature dependence in the current overdrive and write error rate. In the easy cone MTJ, these two performance metrics will reduce as temperature is increased. The results shown in this work demonstrate that the easy-cone MTJ is more suitable to work at high temperatures compared with the pMTJ. Our work provides a guidance for the design of STT-MRAM that is required to operate at high temperatures.

Keywords: spin-transfer torque; easy-cone magnetization; precession switching; current overdrive; write error rate

1. Introduction

Spin-transfer torque magnetic random-access memory (STT-MRAM) with perpendicular magnetization is a promising and emerging technology. Benefiting from its nonvolatility, faster access speed compared with DRAM and higher capacity compared with SRAM, STT-MRAM has become a competitive candidate to replace SRAM and DRAM to realize the universal memory [1,2]. STT-MRAM consists of the transistor and magnetic tunnel junction (MTJ). As shown in Figure 1a, MTJ consists of a nonmagnetic spacer layer and two ferromagnetic layers. One of the ferromagnetic layers is magnetically pinned, named the pinned layer (PL). The magnetization of the other ferromagnetic layer can be changed by the external excitations. This layer is named the free layer (FL). Since the two magnetic layers have the same easy axis, their magnetization, m_{FL} and m_{PL}, can only be parallel or antiparallel. According to the tunneling magnetoresistance effect, the electrical resistance of MTJ is low (high) when m_{FL} and m_{PL} are parallel (antiparallel), which can be used to represent the binary states [3].

In the STT-MRAM, a current flowing through the device generates the STT (indicated by the blue arrows in Figure 1a), which can be used to switch the magnetization of the FL. The magnitude of STT is proportional to the relative angle between m_{FL} and m_{PL} (θ), given as $\tau_{STT}\, m_{FL} \times (m_{FL} \times m_{PL})$, where τ_{STT} is the coefficient dependent on the physical parameters of MTJ [4]. Restricted to the collinear magnetization at equilibrium, the STT is vanished although the current is applied. In consequence, the STT-induced magnetization switching relies on the thermal fluctuations that can break the collinearity of m_{FL} and m_{PL}. Nevertheless, since these fluctuations are random in nature, the switching time is greatly varied and cannot be controlled [5–7]. This stochasticity put forward stringent requirements for the design of the device to achieve an appropriate write error rate [8–10]. Some studies focused on the design of the device structure to increase the relative angle between m_{FL} and m_{PL}. For example, the PL with a tilted easy axis has been employed, but it brought difficulties to the epitaxial growth. Two spin polarizing layers with in-plane and out-of-plane magnetization are also considered. However, it is limited by the complex stacking structure [11–14].

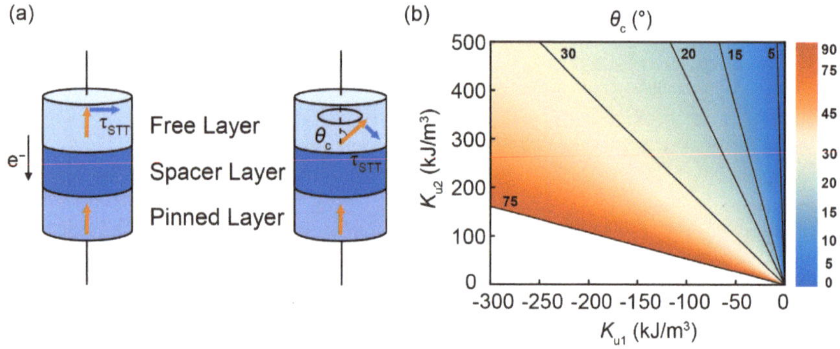

Figure 1. (**a**) Schematic of magnetic tunneling junctions with perpendicular and tilted magnetized FL. The orange arrows represent the magnetization of FL and PL. The blue arrows represent the spin transfer torque exerted on the magnetization. θ_c is defined as the polar angle of easy-cone FL magnetization. (**b**) θ_c as a function of K_{u1} and K_{u2}.

Some analytical and macrospin studies have revealed that the switching characteristics of MTJs can be improved by exploiting a conically magnetized FL [15–18], which can be realized by the second-order magnetic anisotropy. As shown in the right panel of Figure 1a, when the demagnetizing energy is partially or fully balanced by the first-order anisotropy energy, the second-order magnetic anisotropy leads to an easy cone state which has a fixed polar angle (θ_c) with respect to the out-of-plane direction [19–22]. This self-contained misalignment brings advantages such as faster switching, lower switching current density, and better write error rate compared with the pMTJ. However, almost all studies that compare the easy-cone MTJ and pMTJ focus on the room temperature performance. Meanwhile, the temperature studies on the easy-cone MTJ only focus on how to stabilize the easy-cone magnetization under various temperatures or the temperature effect on the TMR ratio [15,19]. The temperature effect on the writing performance of the easy-cone MTJ has not been systematically investigated. It has been reported that there are many issues in the pMTJ as the temperature is varied. For example, at low temperatures, the reduced thermal fluctuations decrease the relative angle between m_{FL} and m_{PL}. The initial STT will thus be decreased and the incubation delay will be increased [23]. However, in the easy-cone MTJ, this angle is mainly determined by the magnetocrystalline anisotropy energy and demagnetizing energy. It will not be constrained by thermal fluctuations. At high temperatures, the thermal stability (Δ) of the pMTJ will be reduced a lot [24]. However, the existence of second-order anisotropy in the easy-cone MTJ increases the energy barrier between the equilibrium magnetization state and the in-plane magnetization state (E_B), which compensates for the reduction in Δ. Δ is closely related to the writing performance of the MTJ. Therefore, it is believed that the writing performance of the easy-cone MTJ is more stable under temperature variations. This makes it more suitable to operate at extreme temperatures.

In this work, the width and length of FL is assumed to be 20 nm and 48 nm, respectively. Previous studies pointed out that the ferromagnets with a lateral size smaller than 80 nm can be described by the macrospin model [25]. Thus, we use the macrospin model to simulate the dynamics of the device. Based on the temperature dependence of physical parameters, we show that the easy-cone MTJ has a larger Δ at high temperatures due to the second-order anisotropy. Meanwhile, Δ changes less significantly in the easy-cone MTJ as the temperature is varied. As a result, the writing performance of the easy-cone MTJ show a better immunity under temperature variations. In addition, our results show that the easy-cone MTJ exhibits a smaller switching current density and a lower write error rate at high temperatures. These results further confirm the remarkable potential of the easy-cone MTJ in memory applications. Meanwhile, our work can stimulate the design of high performance STT-MRAM operating at high temperatures.

2. Methodology

In the easy-cone state, the tilted magnetization is stabilized by the competition of the magnetic anisotropy energy and demagnetization energy. Here, the former consists of both first- and second-order magnetic anisotropy energies. Thus, the energy density of easy-cone FL is given by:

$$\varepsilon = \frac{1}{2}\mu_0 M_{sat}^2(N_{xx}m_x^2 + N_{yy}m_y^2 + N_{zz}m_z^2) + K_{u1}(1 - m_z^2) + K_{u2}(1 - m_z^2)^2 \tag{1}$$

Here, N_{xx}, N_{yy}, and N_{zz} represent demagnetization coefficients in the three dimensions. μ_0 is the vacuum permeability and M_{sat} is the saturation magnetization, which refers to the maximum magnetization that the ferromagnet can reach when it is magnetized by a magnetic field. K_{u1} and K_{u2} are the first- and second-order magnetic anisotropy constants, respectively. The demagnetization energy will generate an in-plane shape anisotropy field, $\mathbf{H}_{IP}$, with the magnitude of M_{sat} $(N_{zz}-N_{xx})$ [26]. Thus, the energy density of FL can be rewritten as:

$$\varepsilon = K_{u1,eff}\left(1 - m_z^2\right) + K_{u2}(1 - m_z^2)^2 \tag{2}$$

where $K_{u1,eff}$ is the effective first-order anisotropy constant with demagnetization energy considered, given by $K_{u1,eff} = K_{u1} - (1/2)\mu_0 M_{sat}{}^2(N_{zz} - N_{xx})$. Figure 1b reveals that the competition of $K_{u1,eff}$ and K_{u2} determines the θ_c. When $K_{u1,eff} < 0$ and $-K_{u2}/K_{u1,eff} > 1/2$ (indicated by the shaded region), the easy-cone state can be stabilized. By minimizing the energy density of the FL, the equilibrium polar angle can be expressed as:

$$\theta_c = \sin^{-1}\sqrt{\frac{-K_{u1,eff}}{2K_{u2}}} \tag{3}$$

For practical applications, θ_c is often required to be smaller than 15° to ensure an appropriate Δ.

In this study, the temperature dependence of physical parameters is included. Here, we mainly consider the dependence of M_{sat}, K_{u1}, and the spin polarization factor P on temperature. They can be modeled by following formulas [27–29]:

$$M_{sat}(T) = M_{sat}(0)[1 - (\frac{T}{T_c})^{3/2}] \tag{4}$$

$$K_{u1}(T) = K(0)[\frac{M_{sat}(T)}{M_{sat}(0)}]^3 \tag{5}$$

$$P(T) = P(0)\left(1 - \beta T^{\frac{3}{2}}\right) \tag{6}$$

where T is the temperature, and $\beta = 2 \times 10^{-5}$ $K^{-3/2}$ is the fitting parameters dependent on materials. T_c = 750 K is the Curie temperature. $M_{sat}(0)$, $K_{u1}(0)$, and $P(0)$ are the values at T = 0 K. Their values are given in Table 1 so that the corresponding values at T = 300 K are consistent with those reported in [30]. The pMTJ and easy-cone MTJ studied in this work differ only in magnetic anisotropy energy. That is, $M_{sat}(0)$ and $P(0)$ of both devices are the same but the $K_{u1}(0)$ is different. As for K_{u2}, previous studies pointed out that the significant K_{u2} does not intrinsically originate from the ferromagnetic interfaces. The spatial fluctuations of film thickness and atomic structure at the interface should be responsible for its emergence [31–33]. Therefore, we exclude the temperature dependence of K_{u2} in this study. As illustrated in Figure 2a–c, all these three parameters decrease as T is increased. The same monotonicity of M_{sat} and K_{u1} makes it difficult to directly derive the changes of Δ as the temperature is varied, and we resort to the numerical calculation. The Δ of the easy-cone state is given by [14]:

$$\Delta = \frac{E_B}{k_B T} = \frac{\left(\varepsilon\left(\theta = \frac{\pi}{2}\right) - \varepsilon(\theta = \theta_c)\right)V}{k_B T} \tag{7}$$

Table 1. Parameters of the system.

Parameter	Unit	Value
$P(0)$	~	0.446
$K_{u1}(0)$ for the pMTJ	J/m^3	1.82×10^6
$K_{u1}(0)$ for the easy-cone MTJ	J/m^3	1.1×10^6
$M_{sat}(0)$	A/m	1.22×10^6
t_{FL}	nm	1.2
l_{FL}	nm	48
w_{FL}	nm	20

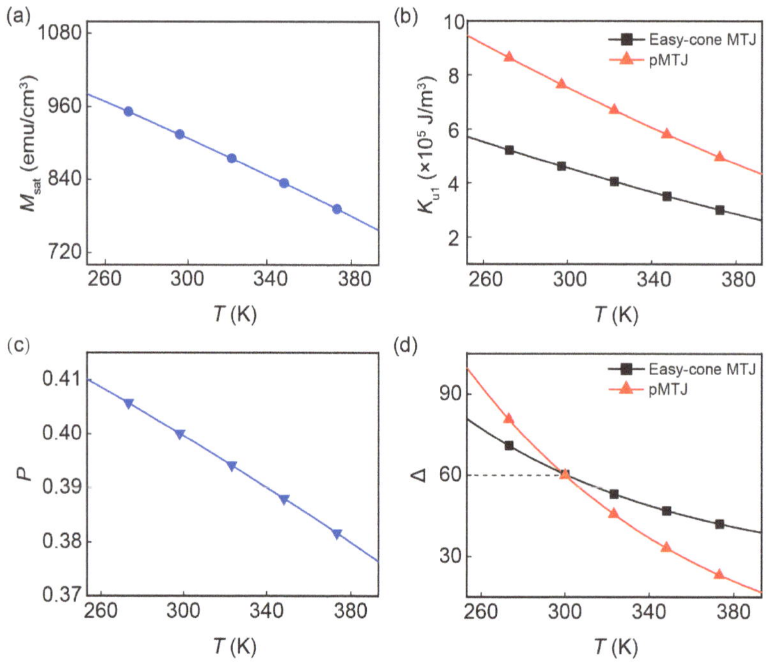

Figure 2. Temperature dependence of physical parameters of MTJ device: (**a**) saturation magnetization M_{sat}, (**b**) first order anisotropy constant K_{u1}, (**c**) spin polarization factor P, and (**d**) thermal stability Δ. Both devices have Δ = 60 at room temperature.

Here, V is the volume of FL and k_B is the Boltzmann constant. Figure 2d shows Δ as a function of T for the pMTJ and easy-cone MTJ. It is important to note that we set Δ = 60 at room temperature (T = 300 K) in both devices for a fair comparison between them. It can be observed that the pMTJ has a higher Δ before the intersection point due to the larger $K_{u1}(0)$. However, it becomes more thermally unstable than the easy-cone MTJ when T is increased. We attribute this to the presence of K_{u2} in the easy-cone MTJ and the faster decreasing K_{u1} in the pMTJ. As T is increased, the invariant K_{u2} in the easy-cone MTJ can produce a stable E_B to slow down the degradation of Δ. In contrast, the rapidly decreased K_{u1} in the pMTJ leads to a sharply reduced E_B. Thus, under high-temperature circumstance, easy-cone MTJ has a superior Δ. This indicates that the easy-cone MTJs have a better performance in the data retention when they are operating at elevated temperatures.

The current driven magnetization dynamics is studied by solving the Landau–Lifshitz–Gilbert–Slonczewski (LLGS) equation $\mathbf{dm}/dt = -\gamma\mathbf{m} \times \mathbf{H}_{\text{eff}} + \alpha\mathbf{m} \times \mathbf{dm}/dt - \gamma\eta\hbar J/(2 e t_{\text{FL}} M_{\text{sat}})\mathbf{m} \times (\mathbf{m} \times \boldsymbol{\sigma}_{\text{STT}})$ [34] with the gyromagnetic γ, the current density J, the reduced Planck constant $\hbar$, the electron charge e, the thickness of FL, t_{FL} = 1.2 nm, the damping constant, α = 0.01, the spin polarization, $\boldsymbol{\sigma}_{\text{STT}}$, and the effective magnetic field, $\mathbf{H}_{\text{eff}}$, which includes the first- and second-order magnetic anisotropy field, the demagnetizing field, and the thermal field. The STT efficiency $\eta = P/[1 + P^2\cos(\theta)]$ is determined by the spin polarization factor and polar angle of FL magnetization.

3. Results and Discussion

To study the temperature effect, we consider the standard operating temperature range of commercial electronic devices, from 273 K to 343 K. To further verify the superiority of the easy-cone MTJ at higher temperatures, we also perform the simulations at the temperature range of 343–373 K. The results show that the easy-cone MTJ is still superior. Therefore, it can be predicted that the advantages of the easy-cone MTJ will become even more apparent above 373 K. When the temperature is smaller than 273 K, it is hard to obtain the easy-cone state since the first-order anisotropy constant is so large that the anisotropy energy cannot be compensated by the demagnetizing energy. Finally, we determined the temperature region 273–373 K. We consider that the temperature affects the writing performance in two ways. One is the temperature dependence of physical parameters, which dominates the variations of the device performance. The other is the influence of the temperature-dependent $\mathbf{H}_{\text{thermal}}$, which arises from the thermal fluctuations. To investigate the intrinsic writing performance, we firstly exclude the $\mathbf{H}_{\text{thermal}}$. To determine the equilibrium state for the easy-cone MTJ, θ_c at various temperatures is calculated using Equation (3), which is shown in Figure 3a. Meanwhile, we determine the θ_c at T = 273 K to T = 373 K using the numerical simulation. The results of the calculation and simulation fit well. It can be seen that the misalignment between the FL and PL will become more pronounced at a higher T. The maximum θ_c in our simulation reaches 19° at T = 373 K. It has been demonstrated that a larger θ_c will lead to lower intrinsic switching current density (J_{sw0}) and switching delay (t_{sw}). J_{sw0} is defined as the critical current density above which STT can overcome the damping and magnetization switching will be initialized. A small J_{sw0} should be promised to ensure a low-power consumption. For the easy-cone MTJ and pMTJ, J_{sw0} is given by Equations (8) and (9), respectively [14,35].

$$J_{\text{sw0,easy-cone}} = \frac{8}{3\sqrt{6}} \frac{\alpha t_{FL}|e|}{\hbar \text{P}} \sqrt{\frac{(K_{u1,eff} + 2K_{u2})^3}{K_{u2}}} \quad (8)$$

$$J_{\text{sw0,PMA}} = \frac{4\alpha t_{FL}|e|}{\hbar \text{P}} K_{u1,eff} \quad (9)$$

As depicted in Figure 3b, J_{sw0} of the easy-cone MTJ and pMTJ will drop as T is increased. In our study, the spin polarization P and $K_{\text{u1,eff}}$ determine J_{sw0} when T varies. The reduced $K_{\text{u1,eff}}$ at high temperatures leads to a lower Δ, and consequently a smaller current density is able to realize magnetization switching. In contrast, the reduced P results in a weaker STT, which calls for additional current density to switch the magnetization [36–38]. Note that $K_{\text{u1,eff}}$ is more sensitive to temperatures, it is the dominated parameter in the variation of J_{sw0}. Thus, the J_{sw0} curves of both easy-cone MTJ and pMTJ show a downward trend. It is worth noting that the decrease in J_{sw0} in pMTJ is steep, while that of the easy-cone MTJ shows a much gradual trend. This is because the variation of $K_{u1,eff}$ in the pMTJ is more significant. Therefore, compared with the easy-cone MTJ, the pMTJ has a much lower J_{sw0} when T is higher than 323 K. However, this is at the cost of its stability under temperature variation. For example, in the range of 273 K to 373 K, J_{sw0} of the pMTJ changes by 56%. This value in the easy-cone MTJ is only 8%. In practical applications, this stability is very attractive since it can promise a wider operating temperature range.

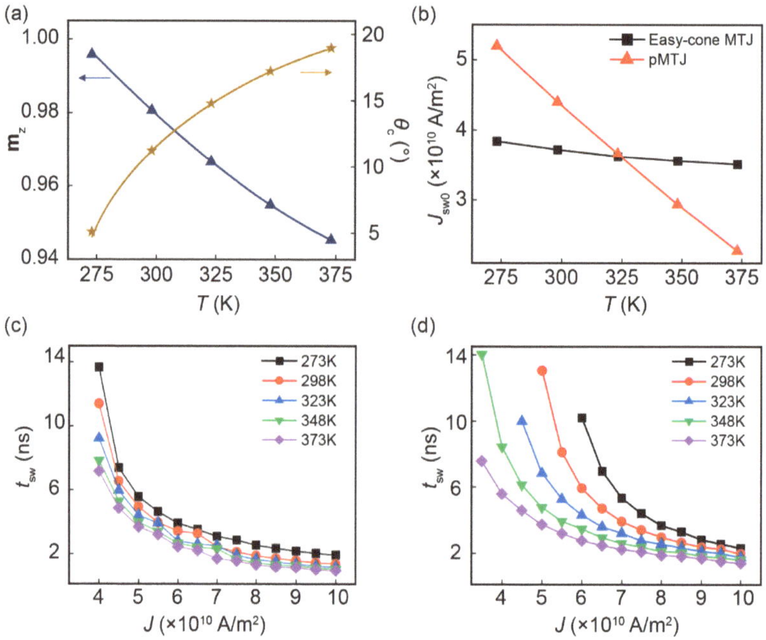

Figure 3. (**a**) The equilibrium $\mathbf{m}_z$ and initial polar angle θ_c as a function of temperature. (**b**) Temperature dependence of the intrinsic switching current density J_{sw0} in the easy-cone MTJ and pMTJ. Switching delay t_{sw} as a function of current density in the (**c**) easy-cone MTJ and (**d**) pMTJ. T = 273 K, 298 K, 323 K, 348 K, and 373 K are represented by square, circle, triangle, down-pointing triangle, and rhombus, respectively.

Next, we examined the switching delay t_{sw}, which characterizes the writing speed of STT-MRAM. Here, t_{sw} is defined as the required time for the magnetic moment switching from $\theta = \theta_c$ to $\theta = 90°$. As illustrated in Figure 3c,d, t_{sw} of both MTJs decreases as T is increased. This stems from the reduced M_{sat} at a high T, which can enhance the STT. In particular, for the easy-cone MTJ, the initial θ_c becomes larger with increasing T, resulting in a larger STT [17]. Such an increase in STT has been confirmed to accelerate the magnetic switching [14]. It is observed that the same as J_{sw0}, for the easy-cone MTJ, the variation of t_{sw} as a function of temperatures is less significant compared with that in the pMTJ. This is more pronounced at moderate current densities 6×10^{10} to 8×10^{10} A/m^2.

At finite temperatures, the thermal fluctuations cannot be ignored, which leads to a stochastic switching. The thermal fluctuations are included as an effective random field:

$$\mathbf{H}_{thermal} = \zeta\sqrt{\frac{2k_B T\alpha}{VM_{sat}(1+\alpha^2)\delta t}} \tag{10}$$

where ζ is a vector with three components that are independent Gaussian random variables, and the time-step δt is 10 ps in our simulation. We then investigate the writing performance of both types of MTJs with pulse duration τ = 1 ns, 2 ns, and 5 ns. When τ < 5 ns, the switching in the conventional pMTJ is in the fast precession regime, where a large current is required for the successful switching [5,6]. As a result, the torque from thermal fluctuations plays a negligible role in the magnetization switching. However, it is important to note that the thermal fluctuations also affect the distribution of initial magnetization, which has a strong impact in the switching current density. Therefore, before studying the switching dynamics, one has to determine θ by taking the thermal fluctuations into account. To capture the thermal distribution of the magnetization, we have solved the LLGS equation with the effective thermal field described by Equation (10). Figure 4a shows the mean of θ with thermal fluctuations considered (θ_{thermal}) as a function of temperatures for the easy-cone MTJ and pMTJ. It can be observed that raising T helps achieve a higher θ_{thermal} for both devices, leading to easier spin transfer switching. In addition, θ_{thermal} of the easy-cone MTJ changes more violently with T. However, thermal fluctuations play different roles in the variations of θ_{thermal} in the two devices. In the pMTJ, a larger T results in an intense thermal fluctuation, and further enhances the precession of the FL magnetization. Therefore, θ_{thermal} will increase accordingly. In contrast, for the easy-cone MTJ, the variation of θ_{thermal} mainly arose from the temperature dependence of the physical parameters. As illustrated in Figure 4b, at high temperatures (T > 298 K), the difference between θ_{thermal} and θ_c, defined as $\Delta\theta$, is within 1°. However, at low temperature T = 273 K, θ_{thermal} was enlarged a lot with respect to the intrinsic θ_c. This is because the effective anisotropy field in the easy-cone MTJ is decreased at low temperatures, which becomes comparable to $\mathbf{H}_{\text{thermal}}$. Three components of the magnetization tend to have the same statistical value under the influence of $\mathbf{H}_{\text{thermal}}$. Thus, thermal fluctuations can help achieve a higher θ_{thermal}.

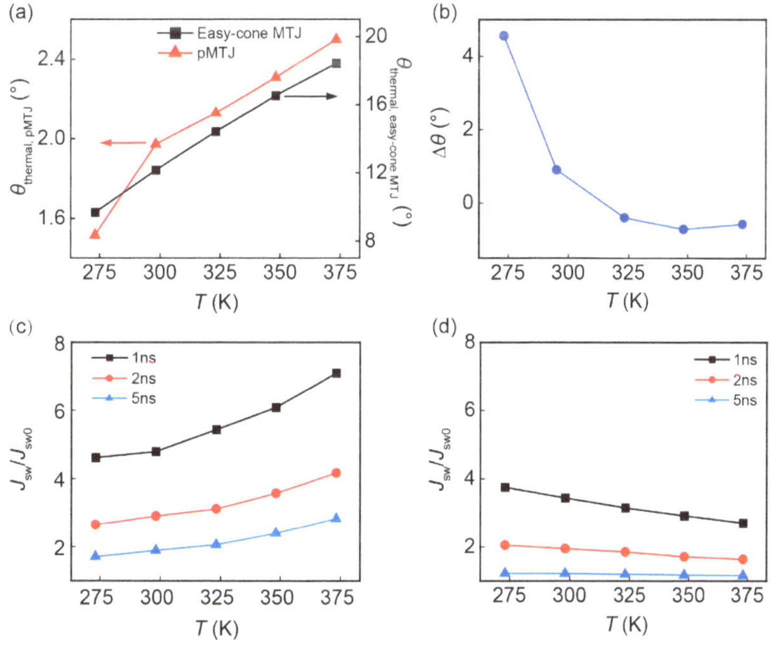

Figure 4. (**a**) θ_{thermal} as a function of temperature for the easy-cone MTJ and pMTJ. (**b**) The difference between θ_c and θ_{thermal} is represented as $\Delta\theta$. The current overdrive as a function of temperature in (**c**) pMTJ and (**d**) easy-cone MTJ.

In the fast precession regime, the required current density for the magnetization switching is several times of J_{sw0}. We define J_{sw} as the current density, at which the switching probability is 50% and investigate how the current overdrive J_{sw}/J_{sw0} is affected when T is varied. The theoretical expression of the overdrive is $1 + [\ln(\pi/2\theta)/J_{sw0}\tau]$, which is determined by the combined effect of θ and J_{sw0} [5]. With increased T, benefiting from the stronger thermal fluctuations and reduced $K_{u1,eff}$, θ will be enlarged, leading to a smaller overdrive. However, this will be compensated by the reduced J_{sw0} at higher T (see Figure 3b). It can be observed in Figure 4c,d that the overdrive of the two devices has an opposite temperature dependence. For the pMTJ, it shows an upward tendency as T is increased, while that of the easy-cone MTJ decreases at elevated T. This is attributed to the larger variations of J_{sw0} compared with the change of θ in the pMTJ. Therefore, J_{sw0} dominates the variations of the overdrive. However, for the easy-cone MTJ, since J_{sw0} barely changes as T is varied, θ dominates the variations of the current overdrive. As mentioned above, θ will be enlarged at higher T. Therefore, the overdrive will decrease as T is raised. In both devices, the variations of the overdrive mainly arise from the temperature dependence of the physical parameters. In contrast, the thermal fluctuations are less important in the fast precession regime. However, when τ is 5 ns, the overdrive of the easy-cone MTJ is almost unchanged as T is varied. It is confirmed that the easy-cone MTJ enters dynamic reversal regime, which is a transitory stage of the fast precession and thermally activated regimes. In this regime, restricted to the short τ, the reduction in overdrive is more gradual. Only when τ is further increased, and the magnetization switching enters thermally activated regime, the overdrive can be further decreased and becomes less than 1 [39]. In conclusion, in the precession regime, the easy-cone MTJ requires lower current density to realize switching than the pMTJ at high temperatures.

The write error rate is the probability of non-switching cases with current applied. In order to ensure the data writing function, the write error rate should be low enough. Figure 5 shows the write error rate as a function of the current density J normalized by J_{sw0} at different τ. An opposite temperature dependence of WER can be observed for the pMTJ and easy-cone MTJ. For the pMTJ, it requires a larger J/J_{sw0} to ensure an appropriate write error rate as T is raised, resulting in the right shift of the write error rate curves. In contrast, for the easy-cone MTJ, the J/J_{sw0} required to achieve the same write error rate becomes smaller at elevated T. For instance, in Figure 5a, the pulse duration is 2 ns and both types of MTJs work in the precession regime. To realize a write error rate of 10^{-3} in the easy-cone MTJ, the required J/J_{sw0} is 2.67 at T = 373 K, and this value is increased to 3.18 at a lower T = 273 K. However, in the pMTJ, the corresponding J/J_{sw0} is 6.31 at T = 373 K and decreased to 3.62 at T = 273 K. It can be observed in Figure 5b that when T changes from 273 K to 373 K, the required J for a write error rate of 10^{-3} at τ = 2 ns will decrease by 22.9% and 23.4% in the easy-cone and pMTJ, respectively. However, as shown in Figure 3b, the J_{sw0} of the pMTJ changes by 56% and this value in the easy-cone MTJ is only 8%. This indicates that although J_{sw0} of the pMTJ has been reduced a lot at elevated T, the required J for an appropriate write error rate cannot benefit from it. In contrast, J in the easy-cone MTJ becomes much smaller when T is increased. This demonstrates that the easy-cone MTJ exhibits a better performance at high temperatures. In addition, it can be observed that the slope of the write error rate curves in both devices is almost unchanged as T is varied. This is attributed to the opposite impact of M_{sat} and P on the write error rate slope. When T is raised, both M_{sat} and P are reduced, and it has been shown that a smaller M_{sat} leads to a larger slope, whereas a reduced P makes the curve more gradual [40]. Since the slope of write error rate curve remains the same, the effect of different T is manifested in the shift of the curves, which is much larger in the pMTJ due to the large variations of J/J_{sw0} required for a specific write error rate. In contrast, there is negligible shifts in the easy-cone MTJ, which is beneficial when the device is required to work stably in a wide temperature range.

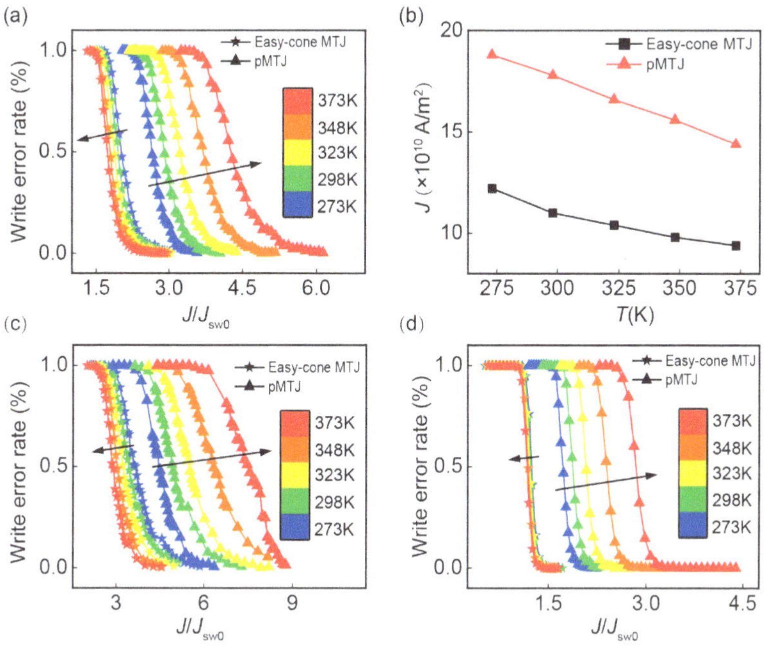

Figure 5. (**a**) Write error rate as a function of current density (normalized by J_{sw0}) with pulse duration τ = 2 ns. The easy-cone MTJ and pMTJ are represented by star and triangle, respectively. The arrow indicates the direction of increased temperature. (**b**) The required current density for the write error rate of 10^{-3} at τ = 2 ns. Write error rate as a function of current density (normalized by J_{sw0}) with pulse duration τ of (**c**) 1 ns and (**d**) 5 ns.

4. Conclusions

In conclusion, we used the macrospin model to investigate the temperature dependence of the writing performance of the easy-cone MTJ and pMTJ. Consistent with the previous studies, at the same temperature, easy-one MTJ has a better writing performance [13,17,24]. In addition, the investigation of the temperature effect revealed that the writing performance of the easy-cone MTJ varies little under temperature variations, leading to its superiority in operating at high temperatures. We first examined the intrinsic switching current density and switching delay. J_{sw0} and t_{sw} of both devices were decreased at high temperatures, and for the pMTJ, they fell even faster. With the effective thermal field introduced, we then performed the stochastic simulations. In the fast precession regime, the easy-cone MTJ and pMTJ had an opposite temperature dependence in the current overdrive J_{sw}/J_{sw0}. At higher temperatures, easy-cone MTJ had a smaller J_{sw}/J_{sw0}. Meanwhile, the write error rate curve shifted in opposite directions with increased temperatures for the two devices. It was also demonstrated that the required current density for an appropriate write error rate is smaller in the easy-cone MTJ. Our work reveals the potential of the easy-cone MTJ-based STT-MRAM in commercial electronic devices. Benefiting from its outstanding performance at high temperatures, the easy-cone MTJ is also fully capable of working in the automotive IC.

Author Contributions: Conceptualization, J.L., Y.Z., and Z.Z.; methodology, J.L., Q.H., Z.Z. and Z.Y.; software, J.L., J.R., Y.X. and Z.Z.; validation, J.L., Z.Y., Q.H., H.L., B.D. and G.L.; formal analysis, J.L., Q.H. and Z.Z.; investigation, J.L. and Z.Y.; resources Z.Z. and Y.Z.; writing—original draft preparation, J.L.; writing—review and editing, J.L., Y.Z., Q.H., Y.Y., Z.Y. and Z.Z. All authors have read and agreed to the published version of the manuscript.

Funding: This research was funded by the National Key R & D Program of China (Grant No. 2022YFB4401700), Shanghai Sailing Program (Grant No. 20YF1430400), and National Natural Science Foundation of China (Grants No. 12104301, 62104130).

Data Availability Statement: Data is available at the correspondence author.

Conflicts of Interest: The authors declare no conflict of interest.

References

1. Fong, X.; Kim, Y.; Venkatesan, R.; Choday, S.H.; Raghunathan, A.; Roy, K.J. Spin-transfer torque memories: Devices, circuits, and systems. *Proc. IEEE* **2016**, *104*, 1449–1488. [CrossRef]
2. Ando, K.; Fujita, S.; Ito, J.; Yuasa, S.; Suzuki, Y.; Nakatani, Y.; Miyazaki, T.; Yoda, H.J. Spin-transfer torque magnetoresistive random-access memory technologies for normally off computing. *J. Appl. Phys.* **2014**, *115*, 172607. [CrossRef]
3. Julliere, M. Tunneling between ferromagnetic films. *Phys. Lett. A* **1975**, *54*, 225–226. [CrossRef]
4. Slonczewskim, J.C. Current-driven excitation of magnetic multilayers. *J. Magn. Magn. Mater.* **1996**, *159*, L1–L7. [CrossRef]
5. Bedau, D.; Liu, H.; Sun, J.; Katine, J.; Fullerton, E.; Mangin, S.; Kent. Spin-transfer pulse switching: From the dynamic to the thermally activated regime. *Appl. Phys. Lett.* **2010**, *97*, 262502. [CrossRef]
6. Huai, Y. Spin-transfer torque MRAM (STT-MRAM): Challenges and prospects. *AAPPS Bull.* **2008**, *18*, 33–40.
7. Joshi, V.K.; Barla, P.; Bhat, S.; Kaushik, B.K. From MTJ device to hybrid CMOS/MTJ circuits: A review. *IEEE Access* **2020**, *8*, 194105–194146. [CrossRef]
8. Shiota, Y.; Nozaki, T.; Tamaru, S.; Yakushiji, K.; Kubota, H.; Fukushima, A.; Yuasa, S.; Suzuki, Y. Evaluation of write error rate for voltage-driven dynamic magnetization switching in magnetic tunnel junctions with perpendicular magnetization. *Appl. Phys. Express* **2015**, *9*, 013001. [CrossRef]
9. Shiota, Y.; Nozaki, T.; Tamaru, S.; Yakushiji, K.; Kubota, H.; Fukushima, A.; Yuasa, S.; Suzuki, Y.J. Reduction in write error rate of voltage-driven dynamic magnetization switching by improving thermal stability factor. *Appl. Phys. Lett.* **2017**, *111*, 022408. [CrossRef]
10. Yamamoto, T.; Nozaki, T.; Imamura, H.; Shiota, Y.; Tamaru, S.; Yakushiji, K.; Kubota, H.; Fukushima, A.; Suzuki, Y.; Yuasa, S. Improvement of write error rate in voltage-driven magnetization switching. *J. Phys. D Appl. Phys.* **2019**, *52*, 164001. [CrossRef]
11. Zhu, X.; Zhu, J.-G. Spin torque and field-driven perpendicular MRAM designs scalable to multi-Gb/chip capacity. *IEEE. Trans. Magn.* **2006**, *42*, 2739–2741. [CrossRef]
12. Zha, C.; Persson, J.; Bonetti, S.; Fang, Y.; Åkerman, J. Pseudo spin valves based on L10 (111)-oriented FePt fixed layers with tilted anisotropy. *Appl. Phys. Lett.* **2009**, *94*, 163108. [CrossRef]
13. He, P.-B.; Li, Z.-D.; Pan, A.-L.; Wan, Q.; Zhang, Q.-L.; Wang, R.-X.; Wang, Y.-G.; Liu, W.-M.; Zou, B.-S. Theory of ferromagnetic resonance in magnetic trilayers with a tilted spin polarizer. *Phys. Rev. B* **2008**, *78*, 054420. [CrossRef]
14. Matsumoto, R.; Arai, H.; Yuasa, S.; Imamura, H. Spin-transfer-torque switching in a spin-valve nanopillar with a conically magnetized free layer. *Appl. Phys. Express* **2015**, *8*, 063007. [CrossRef]
15. Timopheev, A.; Teixeira, B.; Sousa, R.; Aufret, S.; Nguyen, T.; Buda-Prejbeanu, L.; Chshiev, M.; Sobolev, N.; Dieny, B. Inhomogeneous free layer in perpendicular magnetic tunnel junctions and its impact on the effective anisotropies and spin transfer torque switching efficiency. *Phys. Rev. B* **2017**, *96*, 014412. [CrossRef]
16. Matsumoto, R.; Arai, H.; Yuasa, S.; Imamura, H. Theoretical analysis of thermally activated spin-transfer-torque switching in a conically magnetized nanomagnet. *Phys. Rev. B* **2015**, *92*, 140409. [CrossRef]
17. Mojumder, N.N.; Roy, K. Proposal for switching current reduction using reference layer with tilted magnetic anisotropy in magnetic tunnel junctions for spin-transfer torque (STT) MRAM. *IEEE Trans. Electron Devices* **2012**, *59*, 3054–3060. [CrossRef]
18. Matsumoto, R.; Arai, H.; Yuasa, S.; Imamura, H. Efficiency of spin-transfer-torque switching and thermal-stability factor in a spin-valve nanopillar with first-and second-order uniaxial magnetic anisotropies. *Phys. Rev. Appl.* **2017**, *7*, 044005. [CrossRef]
19. Teixeira, B.; Timopheev, A.; Caçoilo, N.; Auffret, S.; Sousa, R.; Dieny, B.; Sobolev, N. Stabilization of the easy-cone magnetic state in free layers of magnetic tunnel junctions. *Phys. Rev. B* **2019**, *100*, 184405. [CrossRef]
20. Timopheev, A.; Sousa, R.; Chshiev, M.; Nguyen, H.; Dieny, B. Second order anisotropy contribution in perpendicular magnetic tunnel junctions. *Sci. Rep.* **2016**, *6*, 26877. [CrossRef]
21. Shaw, J.M.; Nembach, H.T.; Weiler, M.; Silva, T.; Schoen, M.; Sun, J.Z.; Worledge, D.C. Perpendicular magnetic anisotropy and easy-cone state in Ta/Co60Fe20B20/MgO. *IEEE Magn. Lett.* **2015**, *6*, 3500404. [CrossRef]
22. Frömter, R.; Stillrich, H.; Menk, C.; Oepen, H.P. Imaging the cone state of the spin reorientation transition. *Phys. Rev. Lett.* **2008**, *100*, 207202. [CrossRef]
23. Lang, L.; Jiang, Y.; Lu, F.; Wang, C.; Chen, Y.; Kent, A.D.; Ye, L. A low temperature functioning CoFeB/MgO-based perpendicular magnetic tunnel junction for cryogenic nonvolatile random access memory. *Appl. Phys. Lett.* **2020**, *116*, 022409. [CrossRef]
24. Takeuchi, Y.; Sato, H.; Fukami, S.; Matsukura, F.; Ohno, H. Temperature dependence of energy barrier in CoFeB-MgO magnetic tunnel junctions with perpendicular easy axis. *Appl. Phys. Lett.* **2015**, *107*, 152405. [CrossRef]
25. Reichel, V.; Kovács, A.; Kumari, M.; Bereczk-Tompa, É.; Schneck, E.; Diehle, P.; Pósfai, M.; Hirt, A.M.; Duchamp, M.; Dunin-Borkowski, R.E.; et al. Single crystalline superstructured stable single domain magnetite nanoparticles. *Sci. Rep.* **2017**, *7*, 45484. [CrossRef] [PubMed]

26. Matsumoto, R.; Imamura, H.J. Write Error Rate in Bias-Magnetic-Field-Free Voltage-Induced Switching in a Conically Magnetized Free Layer. *Phys. Rev. Appl.* **2022**, *17*, 034063. [CrossRef]
27. Shang, C.H.; Nowak, J.; Jansen, R.; Moodera, J.S. Temperature dependence of magnetoresistance and surface magnetization in ferromagnetic tunnel junctions. *Phys. Rev. B* **1998**, *58*, R2917. [CrossRef]
28. Callen, H.B.; Callen, E. The present status of the temperature dependence of magnetocrystalline anisotropy, and the l(l+1)2 power law. *J. Phys. Chem. Solids* **1966**, *27*, 1271–1285. [CrossRef]
29. Callen, E.R.; Callen, H.B. Static magnetoelastic coupling in cubic crystals. *Phys. Rev.* **1963**, *129*, 578. [CrossRef]
30. Lee, K.-M.; Choi, J.W.; Sok, J.; Min, B.-C. Temperature dependence of the interfacial magnetic anisotropy in W/CoFeB/MgO. *AIP Adv.* **2017**, *7*, 065107. [CrossRef]
31. Ibrahim, F.; Hallal, A.; Kalitsov, A.; Stewart, D.; Dieny, B.; Chshiev, M. Unveiling Temperature-Dependence Mechanisms of Perpendicular Magnetic Anisotropy at Fe/MgO Interfaces. *Phys. Rev. Appl.* **2022**, *17*, 054041. [CrossRef]
32. Heinrich, B.; Monchesky, T.; Urban, R. Role of interfaces in higher order angular terms of magnetic anisotropies: Ultrathin film structures. *J. Magn. Magn. Mater.* **2001**, *236*, 339–346. [CrossRef]
33. Dieny, B.; Vedyayev, A.J.E. Crossover from easy-plane to perpendicular anisotropy in magnetic thin films: Canted anisotropy due to partial coverage or interfacial roughness. *Europhys. Lett.* **1994**, *25*, 723. [CrossRef]
34. Zhu, Z.; Deng, J.; Fong, X.; Liang, G. Voltage-input spintronic oscillator based on competing effect for extended oscillation regions. *J. Appl. Phys.* **2019**, *125*, 183902. [CrossRef]
35. Slonczewski, J.C. Currents, torques, and polarization factors in magnetic tunnel junctions. *Phys. Rev. B* **2005**, *71*, 024411. [CrossRef]
36. Rehm, L.; Wolf, G.; Kardasz, B.; Pinarbasi, M.; Kent, A.D. Sub-nanosecond spin-torque switching of perpendicular magnetic tunnel junction nanopillars at cryogenic temperatures. *Appl. Phys. Lett.* **2019**, *115*, 182404. [CrossRef]
37. Garzón, E.; De Rose, R.; Crupi, F.; Carpentieri, M.; Teman, A.; Lanuzza, M. Simulation analysis of DMTJ-based STT-MRAM operating at cryogenic temperatures. *IEEE Trans. Magn.* **2021**, *57*, 3401406. [CrossRef]
38. Takeuchi, Y.; Enobio, E.C.I.; Jinnai, B.; Sato, H.; Fukami, S.; Ohno, H. Temperature dependence of intrinsic critical current in perpendicular easy axis CoFeB/MgO magnetic tunnel junctions. *Appl. Phys. Lett.* **2021**, *119*, 242403. [CrossRef]
39. Li, Z.; Zhang, S. Thermally assisted magnetization reversal in the presence of a spin-transfer torque. *Phys. Rev. B* **2004**, *69*, 134416. [CrossRef]
40. Xie, Y.; Behin-Aein, B.; Ghosh, A.W. Fokker—Planck study of parameter dependence on write error slope in spin-torque switching. *IEEE Trans. Electron Devices* **2016**, *64*, 319–324. [CrossRef]

nanomaterials

MDPI

Article

A Voltage-Modulated Nanostrip Spin-Wave Filter and Spin Logic Device Thereof

Huihui Li [1], Bowen Dong [1], Qi Hu [1], Yunsen Zhang [1], Guilei Wang [1], Hao Meng [2] and Chao Zhao [1,*]

1 Beijing Superstring Academy of Memory Technology, Beijing 100176, China
2 Changxin Memory Technologies, Inc., Hefei 230601, China
* Correspondence: chao.zhao@bjsamt.org.cn

Abstract: A nanostrip magnonic-crystal waveguide with spatially periodic width modulation can serve as a gigahertz-range spin-wave filter. Compared with the regular constant-width nanostrip, the periodic width modulation creates forbidden bands (band gaps) at the Brillouin zone boundaries due to the spin-wave reflection by the periodic potential owing to the long-range dipolar interactions. Previous works have shown that there is a critical challenge in tuning the band structures of the magnonic-crystal waveguide once it is fabricated. In this work, using micromagnetic simulations, we show that voltage-controlled magnetic anisotropy can effectively tune the band structures of a ferromagnetic–dielectric heterostructural magnonic-crystal waveguide. A uniformly applied voltage of 0.1 V/nm can lead to a significant frequency shift of ~9 GHz. A spin-wave transistor prototype employing such a kind of spin-wave filter is proposed to realize various logical operations. Our results could be significant for future magnonic computing applications.

Keywords: spin wave transistor; spin wave filter; voltage modulated

1. Introduction

Magnonic crystals comprising magnetic microstructures distributed periodically, similar to photonic crystal [1], are core components of complex spin logic devices [2–7]. Typically, the features of spin waves, such as dispersion relations, are characterized by two different magnetic interactions [6,8]: short-range exchange interactions, which are strong, and long-range dipole–dipole interactions, which are relatively weak. In a uniform nanostrip, the influences of dipole–dipole interactions can be simplified as shape anisotropies [9]. Thus, excited spin waves with short wavelengths are mainly governed by exchange interactions. In the presence of periodic modulation, long-range dipolar interaction gives rise to a periodic potential. As a result, spin waves with specific frequencies get reflected and forbidden bands (band gaps) at the Brillouin zone boundaries are formed [10].

In the past few years, various magnonic crystals have been reported. For example, using micromagnetic simulations, bicomponent magnonic crystals composed of two different materials were fabricated [11,12] and investigated [13,14]. A typical implementation of magnonic crystals is the manufacture of periodic modifications based on regular nanostrips [9,15–18]. However, the band structures of magnonic crystals cannot be changed once the periodic modification has been fabricated. A uniform magnetic-field bias applied to magnonic crystals can be used to overcome this limit to tune the band structures [19]. The use of a magnetic field, however, is not desirable for device applications. In this work, the band-structure tuning of magnonic crystals by magnetic anisotropy control using electric fields [10,20–22] was studied.

Voltage-controlled magnetic anisotropy (VCMA) has been widely studied theoretically and experimentally [20,21]. Interfacial perpendicular magnetic anisotropy can be tuned by applying a voltage without a charge current [22]. The VCMA effect has been used to assist the switching of magnetic tunnel junctions [20] and to control the domain wall traps in ferromagnetic nanowires [21]. Voltage-controlled magnetic anisotropy has also been used to construct reconfigurable magnonic crystals [10], spin-wave nanochannels [23], and logic devices [22]. In this paper, we show that band structures can be tuned using a constant voltage applied to the width-modulated nanostrip. Compared to pure voltage-controlled reconfigurable magnonic crystals [10], the voltage used in this work is uniform and the device is easy to fabricate. Moreover, a relatively small voltage (electric field 0.1 V/nm) can cause a significant shift in frequency (~9 GHz).

2. Model and Methods

The width-modulated nanostrip magnonic-crystal waveguide with an ultra-thin cobalt (Co) magnetic layer is illustrated schematically in Figure 1. The entire magnonic-crystal structure is composed of two parts. The left part is a regular nanostrip with a width W_1 = 30 nm, and the remainder is a periodic-width-modulated nanostrip. The in-plane periodic width modulation is characterized by two typical widths, W1 (30 nm) and W2 (18 nm). The width period is P = P1 (18 nm) + P2 (21 nm) = 39 nm, where P1/P2 represents the segment length of the W1/W2-wide nanostrip. Note that the values of P1 and P2 are assumed here to facilitate gigahertz band-gap simulation. In general, the changing P and P1/P ratios will modify the band gaps [9]. A typical period of tens of nanometers will result in band gaps of a few gigahertz for the dipole-exchange spin waves.

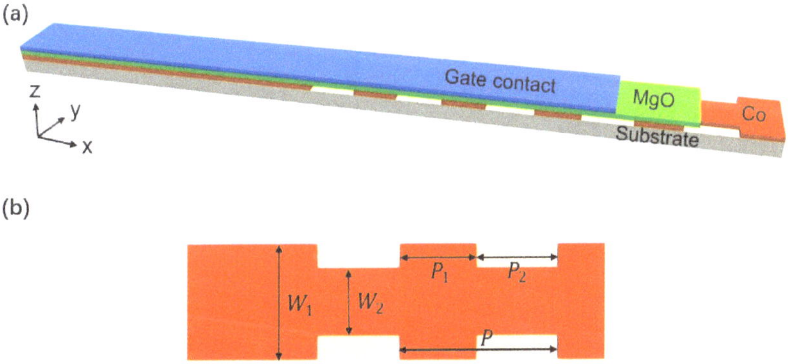

Figure 1. (**a**) Schematic illustration of the width-modulated nanostrip magnonic-crystal waveguides. The film stack is a metal/MgO/Co (1.5 nm)/insulate substrate. The top layer is the gate metal. A voltage was applied on the gate contact to tune the anisotropy. The Co layer is fabricated with periodic width modulations. (**b**) Four parameters characterize the geometry of the Co layer: the width of the nanostrip, W_1, and a second width at the narrow place, W_2; the lengths of the segments with widths W1 and W2 are P_1 and P_2, respectively. In the simulation, we used W_1 = 30 nm, W_2 = 18 nm, P_1 = 18 nm, and P_2 = 21 nm.

We consider the exchange interactions, uniaxial anisotropy, Zeeman energy, and dipole interactions within the system. Therefore, the total free energy of the studied system can be expressed as:

$$E = \int \left[A(\nabla \mathbf{m})^2 - K m_x^2 - \mu_0 \mathbf{m} \cdot \mathbf{h} + E_d\right] dV \quad (1)$$

where A is the exchange constant, $\mathbf{m}$ is the unit vector of the magnetization, K is the uniaxial magnetic anisotropy coefficient, E_d is the demagnetization energy, and $\mathbf{h}$ is the external excitation field. The effect of the voltage-controlled magnetic anisotropy will be taken into account in the uniaxial anisotropy.

The magnetization dynamics is described by the Landau–Lifshitz–Gilbert (LLG) equation:

$$\frac{\partial \mathbf{m}}{\partial t} = -\gamma \mathbf{m} \times \mathbf{H}_{\text{eff}} + \alpha\, \mathbf{m} \times \frac{\partial \mathbf{m}}{\partial t}$$

where $\gamma = 2.11 \times 10^5$ m/(A·s) is the gyromagnetic ratio, $\mathbf{H}_{\text{eff}} = -\frac{1}{\mu_0 M_s}\frac{\delta E}{\delta m}$ is the total effective magnetic field, and α is the Gilbert damping constant. The dispersion relations of the spin-wave excitations are calculated numerically using this equation. We fix $\alpha = 0.01$ [10] in this work for all the simulations.

The ground state of an ultrathin Co film is dependent upon the film's thickness [10]. If the film becomes thinner than the critical thickness, which is typically 0.5–1.0 nm, the ground-state magnetization parallels the out-of-plane axis (z-axis). On the other hand, the magnetization tends to be aligned in-plane if the thickness is larger than the critical thickness. Since VCMA is a pure interface effect, in this study we decided that the thickness of the film would be 1.5 nm with in-plane magnetic anisotropy.

Micromagnetic simulations using the public Object-Oriented Micromagnetic Framework (OOMMF) [24] were performed. The finite-difference methods were used to compute the total free energy in Equation (1). For this simulation, the following typical parameters of Co were chosen [10]: the exchange constant $A = 1.5 \times 10^{-11}$ J/m, the saturation magnetization $M_s = 5.8 \times 10^5$ A/m, and the easy-axis magnetic anisotropy $K = 0$ J/m^3 for an electric field $E = 0$ V/nm. The cell geometry was 1.5 nm × 1.5 nm × 1.5 nm, below the studied material's exchange length.

We computed the band structures of the magnonic crystals by performing a 2D discrete Fourier transform for the temporal and spatial magnetization data, i.e., m_y along the x-axis at $y = 15$ nm, collected every 1 ps for a duration of 8 ns, applying an excitation field to the system. The external field reads [25]:

$$\mathbf{h} = h_0 \text{sinc}(\omega_c(t - t_0))\text{sinc}(k_c(x - x_c)) \sum_{i=1}^{w/dy} \sin\frac{i\pi y}{w}\mathbf{e_y}$$

where $h_0 = 1000$ A/m, sinc(x) = sin(πx)/(πx), $t_0 = 50$ ps, $\omega_c = 120$ GHz, $k_c = 0.1$ nm^{-1}, $x_c = 1500$ nm, w is the width of the wire, and $dy = 1.5$ nm. Both symmetric and antisymmetric modes of spin waves [25] were excited by the above-mentioned magnetic field signal.

3. Results and Discussion

Figure 2 shows the calculated dispersion curves for the spin-wave excitation along the x-axis at $y = 15$ nm. In Figure 2a, spin waves below 4.6 GHz are not allowed to propagate, forming an intrinsic forbidden band. This is because the dispersion relation of spin waves for the nanotrack can be written as:

$$\omega = (2\gamma/\mu_0 M_s)\sqrt{\left(K + Ak^2\right)\left(K + K_\perp + Ak^2\right)}$$

where K and $K_\perp$ are two effective anisotropies. Therefore, the minimum allowed frequency $\omega_c = (2\gamma/\mu_0 M_s)\sqrt{K(K + K_\perp)}$. Owing to the periodic modifications, the lower band gaps Δ_l = 12.6–19.2 GHz emerge at the Brillouin zone (BZ) boundaries, denoted by the blue dashed vertical lines. The BZ boundaries are located at the positions $k_x = (2n + 1)\pi/P$, where n = 0, ±1, ±2, ... and P is the width period of the magnonic crystals. The higher band gaps Δ_h = 37.6–44.2 GHz are found at $k_x = 2n\pi/P$, which are represented by the red dotted vertical lines. Clearly, both the lower and higher band gaps are associated with the one-dimensional translation symmetry of the shape periodicity in the x-direction.

In general, the magnonic-crystal band structures were influenced by W_1 (W_2), P_1 (P_2), and P. In this work, we fixed all of them and studied the influence of applied voltage. We assumed that the strength of the surface anisotropy energy varied linearly with the electric field E, that is, $\Delta K_s = \beta_s E$, where β_s is the magnetoelectric coefficient. Various magnetoelectric coefficients have been reported [10,20,23], and in our simulation we used $\beta_s = 100$ fJ/Vm. Moreover, we assumed that $K_s = 0$ when the applied voltage was zero. Accordingly, an electric field of 0.1 V/nm corresponds to a surficial anisotropy energy of $K_s = 0.1$ mJ/m^2 and thus an effective $\Delta K = K_s/t_c = 6.67 \times 10^4$ J/m^3, where t_c is the thickness of the sample. Figure 2b shows the corresponding dispersion curves when an electric field of 0.1 V/nm was applied. It is clearly shown that the intrinsic forbidden gap was shifted to 13.6 GHz due to the applied voltage (E = 0.1 V/nm). The lower and higher band gaps were changed to Δ_l = 19.9–26.2 GHz and Δ_h = 44.6–51.2 GHz. Therefore, a relatively small voltage can tune the band structures effectively. It is worth mentioning that magnetoelectric coefficients of about 100 fJ/Vm in a CoFeB/MgO [26] system and over 300 fJ/Vm in an iridium-doped Fe/MgO [27] system have been reported. Both material combinations are also promising for VCMA-modulated magnonic crystal applications.

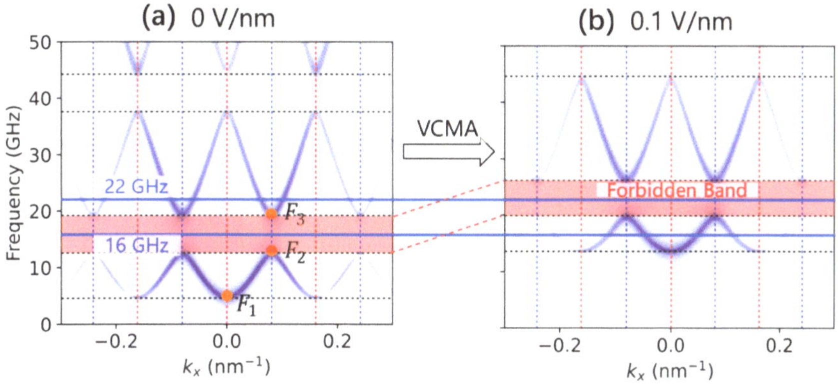

Figure 2. (**a**) Dispersion curves of the spin-wave excitations in the width-modulated nanostrip magnonic crystals. The dispersion curves were calculated using a 2D discrete Fourier transform for the temporal and spatial data at the wire center (y = 15 nm), which were collected every 1 ps after applying an external magnetic field. Lower and higher band gaps of Δ_l = 12.6–19.2 GHz and Δ_h = 37.6–44.2 GHz were observed due to the periodic width modulations. (**b**) The corresponding dispersion curves when applying an electric field of E = 0.1 V/nm. Due to the VCMA-induced anisotropy, the band gaps changed to Δ_l = 19.9–26.2 GHz and Δ_h = 44.6–51.2 GHz. The forbidden band gap is filled in pink. Selected typical frequencies of 16 GHz and 22 GHz are labelled using blue lines to guide the eyes.

Dispersion curves for applied electric fields of 0.05 V/nm and 0.15 V/nm were also calculated. Figure 3 summarizes the typical features of all the simulated dispersion curves. As can be seen in Figure 3a, the three typical frequencies changed nearly linearly with the applied electric fields. Frequency F_1 measured the intrinsic forbidden band, which corresponded to k_x = 0. Frequencies F_2 and F_3 determined the lower band gap, i.e., $\Delta_l = F_2 - F_3$. All of the three typical frequencies increased as the electric field increased. Meanwhile, their incremental amplitudes were different, as shown in Figure 3b. For example, the ΔF for F_1 was the largest, although the value of frequency F_1 was the lowest. The tunabilities of the applied electric fields for F_2 and F_3 were almost the same, indicating a stable forbidden band-gap width.

Here, we give an example to explain the electric-field-modulated spin-wave filter effect in the width-modulated nanostrip waveguide. As shown in Figure 2, the band structures of the spin-wave excitations in the width-modulated magnonic crystals without (left) and with (right) an applied electric field of 0.1 V/nm were compared. Two typical frequencies of 16 GHz and 22 GHz were selected for analysis. Without any electric voltage applied, the spin wave of 16 GHz was in the forbidden band gap and the spin wave of 22 GHz was in the allowed band. That means, intrinsically, that the 16 GHz spin-wave signal could be filtered by this width-modulated nanostrip waveguide. However, after applying an electric field of 0.1 V/nm, the spin wave of 16 GHz was in the allowed band and the spin wave of 22 GHz was in the forbidden band. The spin-wave filtering effect was reversed by the electric field. Therefore, the width-modulated nanostrip magnonic-crystal waveguide can serve as a spin-wave filter [15], and its filtering effect can be tuned by electric fields.

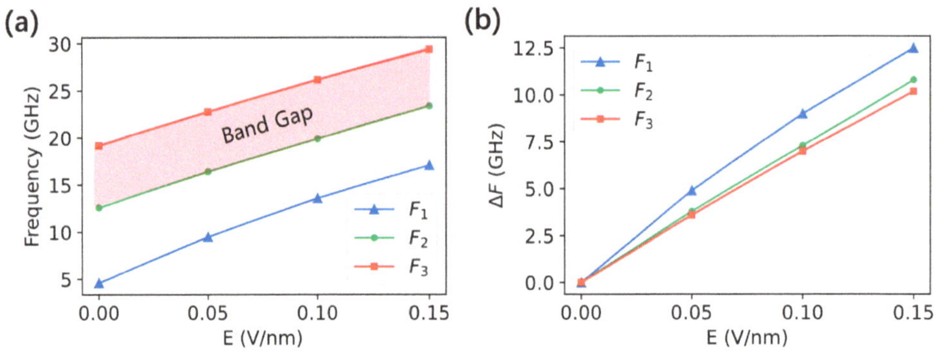

Figure 3. (**a**) The dependence of typical frequencies on applied electric fields. F1 was the lowest allowed frequency which occurred at $k_x = 0$. F2 and F3 determined the lower band gaps at the first BZ boundary, where $k_x = \pi/P$. (**b**) The corresponding incremental frequencies for F_1, F_2, and F_3.

To verify the electric-field-tunable spin-wave filtering effect, we excited spin waves using a *sine* field $\mathbf{h} = h_0\sin(2\pi ft)\mathbf{e}_y$ locally in the region $x \leq 3$ nm, with h_0 = 10,000 A/m. Figure 4a,b show the snapshots of the intrinsic spin-wave propagations without an electric field in a width-modulated nanostrip magnonic-crystal waveguide with excitation frequencies of f = 16 GHz and f = 22 GHz, respectively. The two frequencies selected are typical for the forbidden band and the allowed band, respectively. The red and blue colors represent the z-component of magnetization, m_z. At t = 0.2 ns, spin waves formed in the left regular region of the nanostrip. For the f = 22 GHz case, the spin waves propagated forward, as can be seen at t = 0.3 ns, 0.4 ns, and 0.5 ns. However, as a comparison, spin waves failed to move into the width-modulation region when the excitation frequency f = 16 GHz, as shown in Figure 2, indicating the existence of the forbidden gaps.

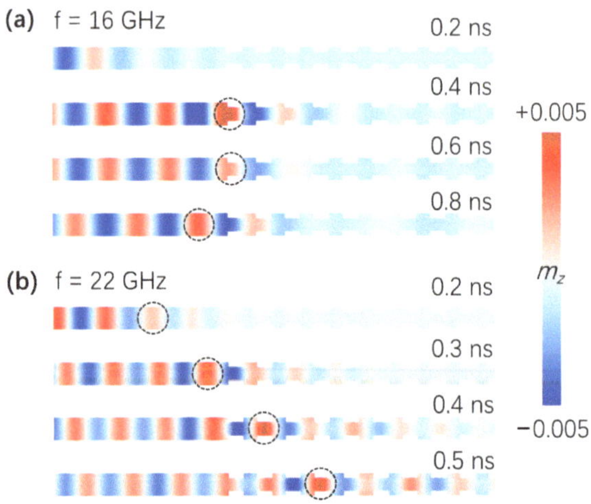

Figure 4. (**a**) Spin-wave propagation along the width-modulated nanostrip magnonic-crystal waveguide without electric field. The spin waves were excited using a sine field $\mathbf{h} = h_0\sin(2\pi ft)\mathbf{e}_y$ locally in the region $x \leq 3$ nm, with h_0 = 10,000 A/m and frequency f = 16 GHz. The dashed circles are shown to guide the eyes. The spin waves could not pass through the narrow part with width modulations. (**b**) As a comparison, the frequency of the excitation field f = 22 GHz was plotted. Clearly, the spin waves propagated even in the presence of width modulations, as the red area moved towards the right with time. A linear color map is used in plotting the m_z.

Figure 5 shows the spin-wave propagation snapshots with an electric field of 0.1 V/nm and excitation frequencies of f = 16 GHz and f = 22 GHz, respectively. In this situation, the frequency f = 22 GHz was located in the forbidden band, while the spin waves with frequency f = 16 GHz were allowed to pass through the width-modulated region, as shown in Figure 2.

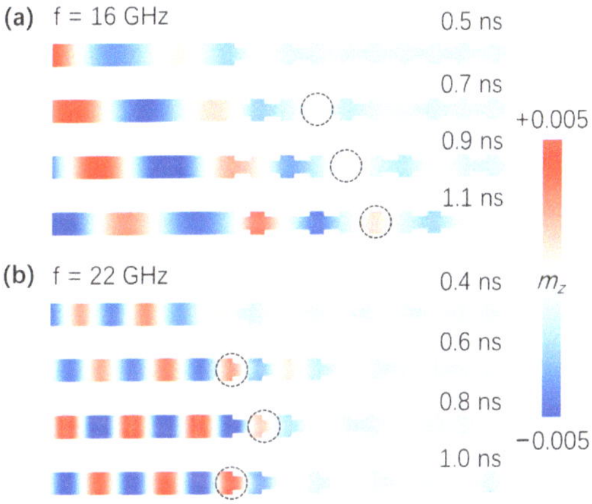

Figure 5. (**a**) Spin-wave propagation along the width-modulated nanostrip with an electric field of 0.1 V/nm. The frequency of the excitation field was f = 16 GHz, which was not in the band gaps. As expected, the spin waves moved forwards along the waveguide. (**b**) The spin waves failed to pass through the width-modulated region when the excitation frequency was f = 22 GHz with the applied electric field.

In Figure 6, an application prototype of a spin-wave transistor using a width-modulated magnonic-crystal spin-wave filter is presented. The regular uniform-width magnetic nanostrip is used as a source terminal (spin-wave injection) and a drain terminal (spin-wave output) for the proposed spin-wave transistor to propagate spin waves. The VCMA system is composed of a metal gate layer, a dielectric layer, and a magnonic-crystal layer that plays a role as a transistor channel to filter the spin waves. The electric voltage applied to the gate can control the spin-wave propagation channel's opening and closing, thus realizing the binary code output of 1 and 0, respectively. The spin-wave transistor proposed here can be used as a fundamental building block for future magnonic computing systems to realize various spin logic operations.

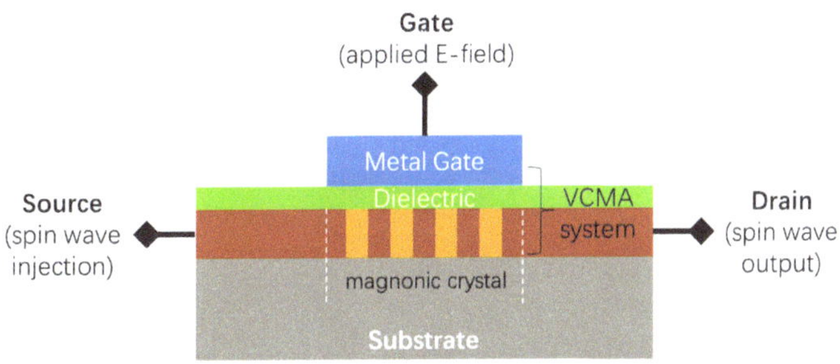

Figure 6. Schematic illustration of a spin-wave transistor proposal based on a VCMA-controlled spin-wave filter using width-modulated magnonic crystals. The VCMA system composed of a metal gate layer, a dielectric layer, and a magnonic-crystal layer was adapted to allow (binary code 1) or forbid (binary code 0) the propagation of spin-wave signals.

4. Conclusions

The voltage-controlled spin-wave filtering behavior in width-modulated nanostrip magnonic-crystal waveguides was studied using micromagnetic simulations. The band structures of the magnonic-crystal waveguide cannot be varied once the periodic modification is fabricated. We have shown that a uniform voltage is sufficient to tune the band structures and that a relatively small voltage with an electric field of 0.1 V/nm can lead to a significant frequency shift (~9 GHz). A spin-wave transistor prototype was finally proposed based on the VCMA-tunable magnonic-crystal spin-wave filter.

Author Contributions: Conceptualization, H.L. and C.Z.; methodology, Y.Z. and G.W.; software, Q.H.; investigation, B.D.; writing—original draft preparation, H.L. and B.D.; writing—review and editing, H.M. All authors have read and agreed to the published version of the manuscript.

Funding: This research received no external funding.

Conflicts of Interest: The authors declare no conflict of interest.

References

1. Vogel, M.; Chumak, A.V.; Waller, E.H.; Langner, T.; Vasyuchka, V.I.; Hillebrands, B.; von Freymann, G. Optically Reconfigurable Magnetic Materials. *Nat. Phys.* **2015**, *11*, 487. [CrossRef]
2. Kruglyak, V.V.; Demokritov, S.O.; Grundler, D. Magnonics. *J. Phys. D Appl. Phys.* **2010**, *43*, 260301. [CrossRef]
3. Kim, S.-K. Micromagnetic Computer Simulations of Spin Waves in Nanometre-Scale Patterned Magnetic Elements. *J. Phys. D Appl. Phys.* **2010**, *43*, 264004. [CrossRef]
4. Tacchi, S.; Duerr, G.; Klos, J.W.; Madami, M.; Neusser, S.; Gubbiotti, G.; Carlotti, G.; Krawczyk, M.; Grundler, D. Forbidden Band Gaps in the Spin-Wave Spectrum of a Two-Dimensional Bicomponent Magnonic Crystal. *Phys. Rev. Lett.* **2012**, *109*, 137202. [CrossRef] [PubMed]
5. Chumak, V.; Serga, A.A.; Hillebrands, B. Magnon Transistor for All-Magnon Data Processing. *Nat. Commun.* **2014**, *5*, 4700. [CrossRef]
6. Chumak, V.; Vasyuchka, V.; Serga, A.; Hillebrands, B. Magnon Spintronics. *Nat. Phys.* **2015**, *11*, 453. [CrossRef]
7. Chumak, V.; Serga, A.A.; Hillebrands, B. Magnonic Crystals for Data Processing. *J. Phys. D Appl. Phys.* **2017**, *50*, 244001. [CrossRef]
8. Kalinikos, A.; Kostylev, M.P.; Kozhus, N.V.; Slavin, A.N. The Dipole-Exchange Spin Wave Spectrum for Anisotropic Ferromagnetic Films with Mixed Exchange Boundary Conditions. *J. Phys. Condens. Matter* **1990**, *2*, 9861. [CrossRef]
9. Lee, K.-S.; Han, D.-S.; Kim, S.-K. Physical Origin and Generic Control of Magnonic Band Gaps of Dipole-Exchange Spin Waves in Width-Modulated Nanostrip Waveguides. *Phys. Rev. Lett.* **2009**, *102*, 127202. [CrossRef]
10. Wang, Q.; Chumak, A.V.; Jin, L.; Zhang, H.; Hillebrands, B.; Zhong, Z. Voltage-Controlled Nanoscale Reconfigurable Magnonic Crystal. *Phys. Rev. B* **2017**, *95*, 134433. [CrossRef]
11. Wang, Z.K.; Zhang, V.L.; Lim, H.S.; Ng, S.C.; Kuok, M.H.; Jain, S.; Adeyeye, A.O. Observation of Frequency Band Gaps in a One-Dimensional Nanostructured Magnonic Crystal. *Appl. Phys. Lett.* **2009**, *94*, 083112. [CrossRef]
12. Gubbiotti, G.; Tacchi, S.; Madami, M.; Carlotti, G.; Yang, Z.; Ding, J.; Adeyeye, A.O.; Kostylev, M. Collective Spin Excitations in Bicomponent Magnonic Crystals Consisting of Bilayer Permalloy/Fe Nanowires. *Phys. Rev. B* **2016**, *93*, 184411. [CrossRef]
13. Ma, F.S.; Lim, H.S.; Wang, Z.K.; Piramanayagam, S.N.; Ng, S.C.; Kuok, M.H. Micromagnetic Study of Spin Wave Propagation in Bicomponent Magnonic Crystal Waveguides. *Appl. Phys. Lett.* **2011**, *98*, 153107. [CrossRef]
14. Ma, F.; Zhou, Y. Interfacial Dzialoshinskii–Moriya Interaction Induced Nonreciprocity of Spin Waves in Magnonic Waveguides. *RSC Adv.* **2014**, *4*, 46454. [CrossRef]
15. Kim, S.-K.; Lee, K.-S.; Han, D.-S. A Gigahertz-Range Spin-Wave Filter Composed of Width-Modulated Nanostrip Magnonic-Crystal Waveguides. *Appl. Phys. Lett.* **2009**, *95*, 082507. [CrossRef]
16. Chumak, V.; Pirro, P.; Serga, A.A.; Kostylev, M.P.; Stamps, R.L.; Schultheiss, H.; Vogt, K.; Hermsdoerfer, S.J.; Laegel, B.; Beck, P.A.; et al. Spin-Wave Propagation in a Microstructured Magnonic Crystal. *Appl. Phys. Lett.* **2009**, *95*, 262508. [CrossRef]
17. Obry; Pirro, P.; Brächer, T.; Chumak, A.V.; Osten, J.; Ciubotaru, F.; Serga, A.A.; Fassbender, J.; Hillebrands, B. A Micro-Structured Ion-Implanted Magnonic Crystal. *Appl. Phys. Lett.* **2013**, *102*, 202403. [CrossRef]
18. Frey, P.; Nikitin, A.A.; Bozhko, D.A.; Bunyaev, S.A.; Kakazei, G.N.; Ustinov, A.B.; Kalinikos, B.A.; Ciubotaru, F.; Chumak, A.V.; Wang, Q.; et al. Reflection-Less Width-Modulated Magnonic Crystal. *Commun. Phys.* **2020**, *3*, 17. [CrossRef]
19. Nikitin, A.; Ustinov, A.B.; Semenov, A.A.; Chumak, A.V.; Serga, A.A.; Vasyuchka, V.I.; Lähderanta, E.; Kalinikos, B.A.; Hillebrands, B. A Spin-Wave Logic Gate Based on a Width-Modulated Dynamic Magnonic Crystal. *Appl. Phys. Lett.* **2015**, *106*, 102405. [CrossRef]
20. Wang, W.-G.; Li, M.; Hageman, S.; Chien, C.L. Electric-Field-Assisted Switching in Magnetic Tunnel Junctions. *Nat. Mater.* **2011**, *11*, 64. [CrossRef]
21. Bauer, U.; Emori, S.; Beach, G.S.D. Voltage-Controlled Domain Wall Traps in Ferromagnetic Nanowires. *Nat. Nanotechnol.* **2013**, *8*, 411. [CrossRef] [PubMed]

22. Rana; Otani, Y. Voltage-Controlled Reconfigurable Spin-Wave Nanochannels and Logic Devices. *Phys. Rev. Appl.* **2018**, *9*, 014033. [CrossRef]
23. Choudhury, S.; Chaurasiya, A.K.; Mondal, A.K.; Rana, B.; Miura, K.; Takahashi, H.; Otani, Y.; Barman, A. Voltage Controlled On-Demand Magnonic Nanochannels. *Sci. Adv.* **2020**, *6*, eaba5457. [CrossRef] [PubMed]
24. Donahue, M.J.; Porter, D.G. OOMMFUser's Guide, Version 1.2a5. Available online: https://math.nist.gov/oommf/doc/userguide12a5/userguide12a5_20120928.pdf (accessed on 28 September 2012).
25. Kumar; Sabareesan, P.; Wang, W.; Fangohr, H.; Barman, A. Effect of Hole Shape on Spin-Wave Band Structure in One-Dimensional Magnonic Antidot Waveguide. *J. Appl. Phys.* **2013**, *114*, 023910. [CrossRef]
26. Li, X.; Fitzell, K.; Wu, D.; Karaba, C.T.; Buditama, A.; Yu, G.; Wong, K.L.; Altieri, N.; Grezes, C.; Kioussis, N.; et al. Enhancement of Voltage-Controlled Magnetic Anisotropy through Precise Control of Mg Insertion Thickness at CoFeB | MgO Interface. *Appl. Phys. Lett.* **2017**, *110*, 052401. [CrossRef]
27. Nozaki, T.; Kozioł-Rachwał, A.; Tsujikawa, M.; Shiota, Y.; Xu, X.; Ohkubo, T.; Tsukahara, T.; Miwa, S.; Suzuki, M.; Tamaru, S.; et al. Highly Efficient Voltage Control of Spin and Enhanced Interfacial Perpendicular Magnetic Anisotropy in Iridium-Doped Fe/Mgo Magnetic Tunnel Junctions. *NPG Asia Mater.* **2017**, *9*, e451. [CrossRef]

nanomaterials

MDPI

Article

Improved Endurance of Ferroelectric $Hf_{0.5}Zr_{0.5}O_2$ Using Laminated-Structure Interlayer

Meiwen Chen [1,2], Shuxian Lv [1,2], Boping Wang [1,2], Pengfei Jiang [1,2], Yuanxiang Chen [1,2], Yaxin Ding [1,2], Yuan Wang [1,2], Yuting Chen [1,2] and Yan Wang [1,2,*]

1 Key Laboratory of Microelectronics Devices and Integrated Technology, Institute of Microelectronics of Chinese Academy of Sciences, Beijing 100029, China; chenmeiwen@ime.ac.cn (M.C.); lvshuxian@ime.ac.cn (S.L.); wangboping@ime.ac.cn (B.W.); jiangpengfei@ime.ac.cn (P.J.); chenyuanxiang@ime.ac.cn (Y.C.); dingyaxin@ime.ac.cn (Y.D.); wangyuan@ime.ac.cn (Y.W.); chenyuting@ime.ac.cn (Y.C.)

2 University of Chinese Academy of Sciences, Beijing 100049, China

* Correspondence: wangyan@ime.ac.cn

Abstract: In this article, the endurance characteristic of the TiN/HZO/TiN capacitor was improved by the laminated structure of a ferroelectric $Hf_{0.5}Zr_{0.5}O_2$ thin film. Altering the HZO deposition ratio, the laminated-structure interlayer was formed in the middle of the HZO film. Although small remanent polarization reduction was observed in the capacitor with a laminated structure, the endurance characteristic was improved by two orders of magnitude (from 10^6 to 10^8 cycles). Moreover, the leakage current of the TiN/HZO/TiN capacitor with the laminated-structure interlayer was reduced by one order of magnitude. The reliability enhancement was proved by the Time-Dependent Dielectric Breakdown (TDDB) test, and the optimization results were attributed to the migration inhibition and nonuniform distribution of oxygen vacancies. Without additional materials and a complicated process, the laminated-structure method provides a feasible strategy for improving HZO device reliability.

Keywords: ferroelectrics; ZrO_2–HfO_2; interlayer; laminated structure; endurance; reliability; oxygen vacancy

1. Introduction

With the emergence of ferroelectricity in doped HfO_2 films, HfO_2-based ferroelectric memory has been expected to be a competitive candidate as a next-generation, nonvolatile memory with excellent scalability, low power consumption, fast speed and complementary metal–oxide–semiconductor (CMOS) compatibility [1–8]. The polycrystalline-fluorite-structure HfO_2 usually behaves as a monoclinic phase at room temperature. With the temperature increasing, there will be transitions among monoclinic, tetragonal and cubic phases [9]. Previous studies revealed that the asymmetric orthorhombic phase with a $Pca2_1$ space group is the origin of ferroelectricity [1]. As the thickness decreases through the atomic layer deposition (ALD) process, the tetragonal phases can be more stable at normal temperatures and pressures [10]. For HfO_2-doped ferroelectric devices, dopants (e.g.: Si, Y, Zr, La, Al, Sr) are used to evoke and stabilize the ferroelectric phase [1,11–15]. Zr dopant is the most commonly used because of the wide doping ratio range and low crystallization temperature which is more compatible with back-end-of-line (BEOL) fabrication [16]. Usually, HZO solid solution films with a HfO_2 and ZrO_2 ratio of 1:1 ($Hf_{0.5}Zr_{0.5}O_2$) are chosen to achieve the optimal ferroelectricity [17,18].

However, the endurance characteristic of the HfO_2-based ferroelectric memory is still an issue to be taken seriously. Because of the destructive reading, the ferroelectric random-access memories are rewritten after each read operation. So, high endurance is necessary for ferroelectric memories. In recent reports, the endurance characteristic of

HfO_2-based ferroelectric devices was lower than that of perovskite-based devices [2,19]. Several methods have been proposed to achieve higher endurance, such as stress control, dopants, temperature regulation and grain boundary interruption. For stress control, the compressive stress using Cu capping applied on HZO thin films was found to achieve high endurance but smaller remanent polarization [20]. Cao et al. found that Ru electrodes can exhibit a lower leakage current and higher breakdown voltage compared with TiN electrodes, which resulted in a higher coercive field [21]. For dopants, Walke et al. demonstrated a high endurance but prolonged wake-up in La and Y doped HZO ferroelectric thin films [22]. For temperature regulation, Choi et al. reported that the different endurance characteristics were observed under different annealing and deposition temperatures [23]. For grain boundary interruption, the most common structure is HZO/Al_2O_3/HZO. Xu et al. reported that the grain boundaries penetrating HZO ferroelectric thin films were interrupted by inserting an Al_2O_3 layer, and the leakage current was reduced, which could enhance the reliability of HZO devices [24]. However, the optimal grain size is limited by the Al_2O_3 interlayer, resulting in small remanent polarization.

Recently, some researchers have focused on the study of thin film deposition methods to adjust the properties of devices. In 2016, Lu et al. reported the induced ferroelectricity of a ZrO_2/HfO_2 bilayer, which provided the basis for the emergence of the laminated structure [25]. Subsequently, Weeks et al. reported a laminated structure (1 nm HfO_2/1 nm ZrO_2) $\times$ 4 which showed the promising ferroelectricity [26]. However, the endurance properties have been less studied in these previous reports. In 2022, Liang et al. studied the performance of a superlattice structure composed of ZrO_2/HfO_2 at different annealing temperatures using a ZrO_2 layer as the starting layer, and reached high endurance performance but low remanent polarization [27]. In 2023, Lehninger et al. studied the performance of HfO_2/ZrO_2 superlattices, which showed an obvious wake-up effect, low coercive field, enhanced polarization and high temperature reliability. However, the endurance reached 10^7 cycles [28]. It is reported that the HfO_2-starting laminated structure exhibited a higher remanent polarization, and optimal remanent polarization was achieved in a thicker sublayer of about 1 nm [28,29]. The ZrO_2-starting laminated structure showed a lower remanent polarization, and the remanent polarization decreased with the deposition cycles increasing [30]. However, it also reported that the ZrO_2 nucleation layer could stabilize the remanent polarization of HZO ferroelectric thin films during field cycling [31]. These previous studies demonstrated the potential of a laminated structure and superlattice structure for improving the ferroelectric and endurance performance.

In this article, we propose a new laminated structure to improve the endurance performance and maintain high remanent polarization. The deposition ratio of the ZrO_2 and HfO_2 layers was adjusted with the ZrO_2 as the starting layer, and the laminated-structure interlayer was formed in the film. The leakage current was reduced by one order of magnitude. The endurance was improved by two orders of magnitude. Furthermore, the breakdown voltage and Time-Dependent Dielectric Breakdown (TDDB) reliability were also enhanced in the TiN/HZO/TiN capacitor with the laminated-structure interlayer. Combined with the migration inhibition and nonuniform distribution of oxygen vacancies, the possible physical mechanisms of laminated structure device performance were analyzed.

2. Materials and Methods

Figure 1a–c illustrate the structures of three types of TiN/HZO/TiN capacitors. The TiN/HZO/TiN capacitors were fabricated on Si/SiO_2 substrates. The 40-nm-thick TiN bottom electrodes were deposited by ion beam sputtering. The $Hf_{0.5}Zr_{0.5}O_2$ ferroelectric thin films were deposited by ALD at 280 °C using $Hf[N(C_2H_5)CH_3]_4$ (Tetrakis(ethylmethylamido) hafnium, TEMAHf), $Zr[N(C_2H_5)CH_3]_4$ (Tetrakis(ethylmethylamido) zirconium, TEMAZr) and H_2O as an Hf precursor, Zr precursor and oxygen source, respectively. Using ZrO_2 as the starting layer, the ZrO_2 and HfO_2 layers were deposited at the same rate of ~1 Å/cycle. The capacitor HZO1/1 was deposited by alternating 1 cycle of ZrO_2 and 1 cycle of HfO_2. Each unit consisted of a ZrO_2/HfO_2 bilayer, and the HZO1/1 thin film comprised 60 units with a thickness of 12 nm, as shown in Figure 1a. The capacitor HZO5/5 was deposited by alternating 5 cycles of ZrO_2 and 5 cycles of HfO_2, which was composed of 12 units. In addition, the thickness was 12 nm, as shown in Figure 1b. In the case of capacitor HZO1/5/1, first, a 5-nm-thick $Hf_{0.5}Zr_{0.5}O_2$ film was deposited by 1 cycle of ZrO_2 and 1 cycle of HfO_2, which was composed of 25 units. Second, the deposition ratio of the ZrO_2 and HfO_2 layers was adjusted, and a 2-nm-thick laminated-structure interlayer was deposited. The interlayer was formed by repeating 5 cycles of ZrO_2 and 5 cycles of HfO_2 and consisted of 2 units. Third, the first deposition step was repeated for a 5 nm thickness. The total thickness of the HZO1/5/1 capacitor was 12 nm, as shown in Figure 1c. Following the deposition of the HZO thin film, the electrodes were patterned through photolithography on HZO. Subsequently, the 40-nm-thick TiN top electrodes were deposited by sputtering. Finally, all the samples were annealed by rapid thermal annealing (RTA) in a N_2 atmosphere. The annealing process involved ramping up the temperature at a rate of 8.4 °C/s for 60 s, and the samples were held at a temperature of 500 °C for 30 s.

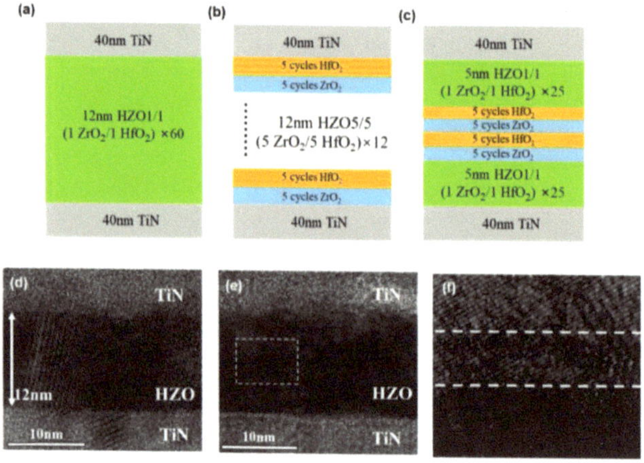

Figure 1. Schematic diagrams of (**a**) the capacitor HZO1/1, (**b**) the capacitor HZO5/5 and (**c**) the capacitor HZO1/5/1. Cross-sectional TEM images of (**d**) the capacitor HZO1/1 and (**e**) the capacitor HZO1/5/1. (**f**): Magnified image extracted from the white box in (**e**).

The electrical characteristics of the capacitors, including the current density–voltage (J–V), dielectric constant–voltage (k–V), polarization–voltage (P–V), and current–time (I–t) curves, were measured using a Keysight B1500 semiconductor parameter analyzer or Radiant Workstation ferroelectric analyzer.

3. Results

3.1. Basic Electrical Properties

Figure 1d–f show the cross-sectional transmission electron microscopy (TEM) images of capacitors HZO1/1 and HZO1/5/1. The thickness of 12 nm was verified in two capacitors by the TEM images. The interfaces between the TiN electrodes and ferroelectric films were also clearly observed, and the bottom interfaces were clearer due to the sputtering process compared to the top interfaces. In HZO1/1, the lattice arrangement was orderly, and in HZO1/5/1, the laminated-structure interlayer was observed, as seen in Figure 1f,e.

Figure 2a,b show the P–V and ε_r–V characteristics of capacitors HZO1/1, HZO5/5 and HZO1/5/1. The P–V loops were obtained after a wake-up cycling of 10^3 with a pulse voltage of 3 V at 1 kHz. As shown in the P–V loops of Figure 2a, the remanent polarizations (P_r) of three capacitors were about 25.37 μC/cm^2, 13.18 μC/cm^2 and 21.87 μC/cm^2, respectively. The remanent polarization was slightly reduced in HZO1/5/1 compared with HZO1/1, while the remanent polarization was significantly reduced in HZO5/5.

The ε_r–V curves of ferroelectric films exhibited typical butterfly-shaped loops at a double sweep voltage from −3 V to 3 V at 100 kHz. The dielectric constants of three capacitors were 33.66, 29.26 and 30.18, respectively. Previous studies have reported the existence of the tetragonal phase and FE orthorhombic phase with a high ε_r value in HfO_2-based ferroelectric films (T: ε_r = 35–40, O: ε_r = 25–30),and the ε_r value of the monoclinic phase is much lower (ε_r = 15–20) [32]. The lower ε_r value of HZO5/5 and HZO1/5/1 compared to HZO1/1 may reveal a higher M-phase fraction in the HZO5/5 and HZO1/5/1 capacitors.

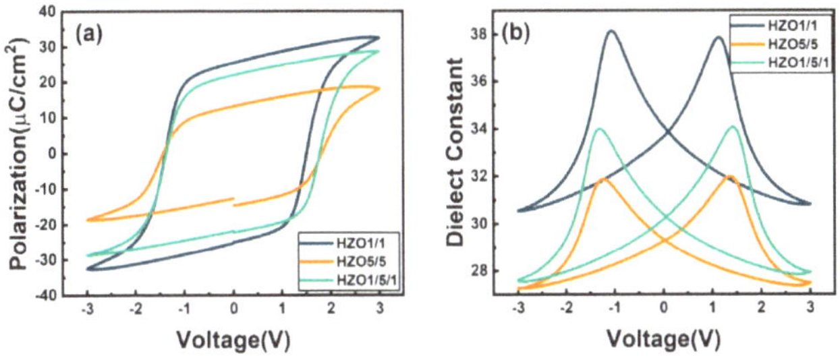

Figure 2. (**a**) P–V loops and (**b**) ε_r–V curves of HZO1/1, HZO5/5 and HZO1/5/1 capacitors.

Figure 3 shows the initial J–V curves of three capacitors. The initial leakage current was measured under a sweep voltage from −3 V to 3 V. The initial current densities of HZO1/1 and HZO5/5 were similar. However, compared with HZO1/1, the initial leakage current of the HZO1/5/1 capacitor was reduced by one order of magnitude (from 2.18×10^{-8} A/cm^2 to 1.51×10^{-9} A/cm^2) at a voltage ±1 V. The leakage current was effectively inhibited in HZO1/5/1. The interfaces of the laminated-structure interlayer could prevent the conduction current path [33,34].

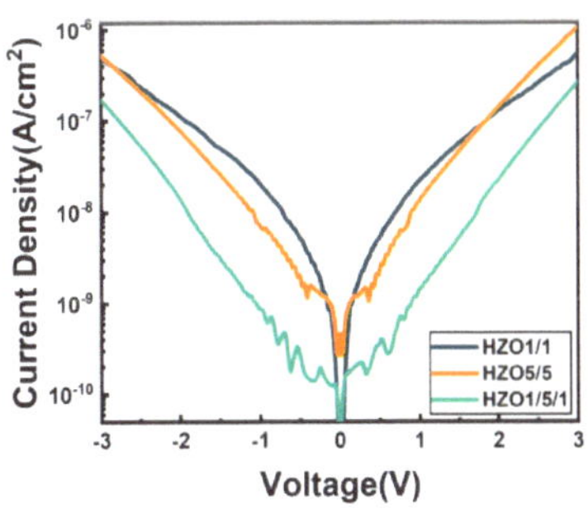

Figure 3. Initial J–V curves of HZO1/1, HZO5/5 and HZO1/5/1 capacitors.

3.2. Endurance and Time-Dependent Dielectric Breakdown Properties

Figure 4 presents the endurance performance of three capacitors in terms of the leakage current and the remanent polarization. The cycling test was taken under a triangular pulse with pulse amplitude of ±3 V and pulse frequency of 0.1 MHz. As shown in Figure 4a, the leakage current of HZO5/5 slightly increased from 1.07×10^{-6} A/cm^2 to 1.35×10^{-6} A/cm^2. However, as the pulse cycles increased, the leakage currents obviously increased in the HZO1/1 and HZO1/5/1 capacitors. For HZO1/1, the leakage current sharply increased to 10^{-3} A/cm^2 after 10^6 cycles. However, the leakage current increased from 2.51×10^{-7} A/cm^2 to 2.36×10^{-6} A/cm^2 in the HZO1/5/1 after 10^8 cycles. It can be inferred that the leakage currents during the cycling could be significantly suppressed in the laminated structure, resulting in an improved endurance performance. The remanent polarization of the three capacitors was measured with the cycling test under an electric field of 2.5 MV/cm, as shown in Figure 4b. A hard breakdown was observed in the HZO1/1 when the electric field cycling exceeds ~10^6 cycles, which is not observed in the HZO5/5 and HZO1/5/1 even up to 10^8 cycles. The endurance of HZO5/5 and HZO1/5/1 could be improved by two orders of magnitude compared with HZO1/1. The $2P_r$ value of HZO5/5 was slightly reduced from 25.9 μC/cm^2 to 20.9 μC/cm^2, and the $2P_r$ value of HZO1/5/1 was reduced from 43.8 μC/cm^2 to 21.6 μC/cm^2 up to 10^8 cycles, but this was still sufficient for memory operation under high cycles.

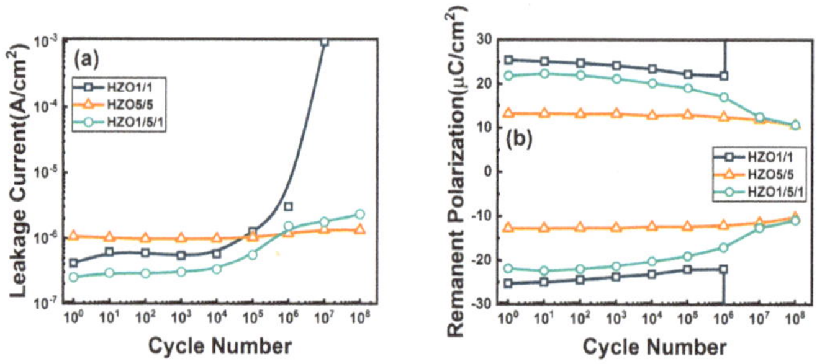

Figure 4. The endurance characteristics of (**a**) leakage currents and (**b**) remanent polarization under cycling pulse with amplitude of ±3 V at 0.1 MHz for HZO1/1, HZO5/5 and HZO1/5/1 capacitors.

Figure 5 shows the statistical chart of the maximum endurance and P_r max characteristics of HZO1/1 and HZO1/5/1, where ten measurements were taken for each box to display the sample-to-sample variation. The maximum endurance cycles ranged from 7×10^5 to 5×10^6 for HZO1/1 and 10^8 to 5×10^8 for HZO1/5/1, indicating that the endurance of HZO1/5/1 was improved by more than two orders of magnitude due to the laminated-structure interlayer. The range of the P_r max value was 22.3 μC/cm^2 to 25 μC/cm^2 for HZO1/1 and 18 μC/cm^2 to 21.9 μC/cm^2 for HZO1/5/1. High remanent polarizations were exhibited in both the HZO1/1 and HZO1/5/1 capacitors.

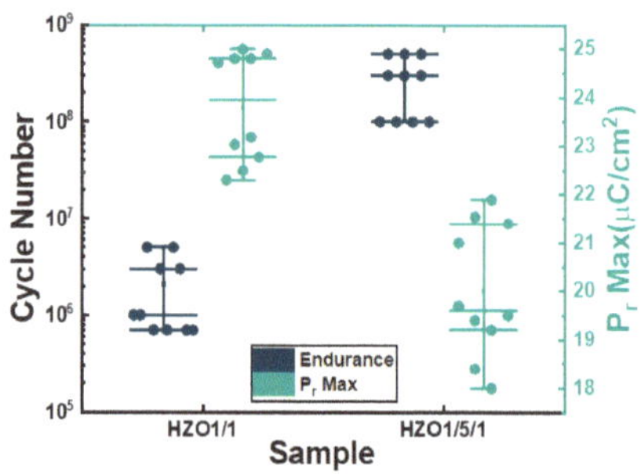

Figure 5. Box chart for endurance and P_r max characteristics of HZO1/1 and HZO1/5/1 sample-to-sample variation. Each box plot consists of ten measurements.

Figures 4 and 5 demonstrate that the endurance performance of HZO capacitors was effectively improved by the laminated-structure interlayer. The suppression of the leakage current in the HZO1/5/1 was attributed to the prevention of the conduction current path by the interfaces of the laminated-structure interlayer. The high remanent polarization values observed in the improved capacitor HZO1/5/1 suggest its application potential in memory devices.

To further investigate the long-term reliability of the HZO1/1 and HZO1/5/1 capacitor, the Time-Dependent Dielectric Breakdown (TDDB) test was utilized with the Constant Voltage Stress (CVS) method. Figure 6a,b show the I–t curves of HZO1/1 and HZO1/5/1 using CVS at three voltages with the area of 1.6×10^{-5} cm^2, and the failure current threshold was set to 1 mA. The hard breakdown could be clearly observed with the top and bottom metal electrodes. Therefore, the breakdown time (T_{BD}) was extracted directly. The set voltages of HZO1/1 were 3.1 V, 3.2 V and 3.3 V. For the similar range of the T_{BD} of two capacitors under CVS, the set voltages of HZO1/5/1 were higher by 0.4 V than those of the HZO1/1. As the DC stress voltage increased, the T_{BD} significantly decreasd. This was due to the generation of defects under CVS, which could form conductive paths through the top and bottom electrodes, leading to breakdown [35]. When the leakage current was increased tenfold, the statistics for T_{BD} of the capacitors at three voltages could be established by the Weibull distribution [36].

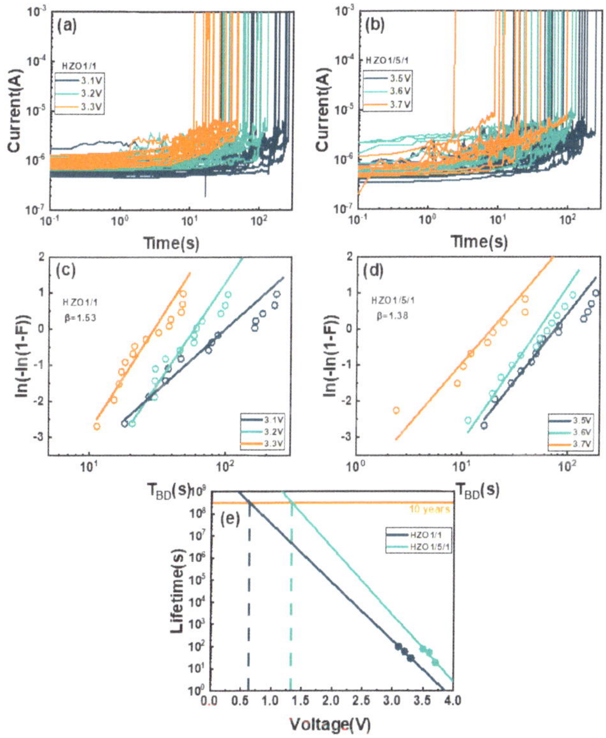

Figure 6. I–t curves of (**a**) the capacitor HZO1/1 and (**b**) the capacitor HZO1/5/1 at three voltages. Weibull distribution of TBD of (**c**) the capacitor HZO1/1 and (**d**) the capacitor HZO1/5/1. (**e**) Ten-year lifetime prediction at 63.2% failure of HZO1/1 and HZO1/5/1 capacitors.

The Weibull plot was applied with cumulative density probability, and the maximum likelihood method was used to fit the data under DC stress, as shown in Figure 6c,d. The cumulative density function in the Weibull plot was given by:

$$W(x) = Ln(-Ln(1 - F(x))) = \beta Ln(x/\alpha) \quad (1)$$

where x is the T_{BD}, α is the scale-factor, and β is the shape-factor.

The relationship between the operating voltage and lifetime could be obtained by the statistics of the 63.2% failure points in the Weibull distribution, and linearly fitting and extrapolating the DC voltage under the failure point. Figure 6e shows the ten-year lifetime of the two capacitors. For the ten-year lifetime prediction at 63% failure, the operating voltage for HZO1/1 was only 0.64 V, which was much smaller than its coercive voltage. This means that at the operating voltage, the lifetime of HZO1/1 cannot reach ten years.

However, the operating voltage for HZO1/5/1 was 1.33 V, which was higher than that for HZO1/1 by 0.6 V. The lifetime of HZO1/5/1 was effectively enhanced compared with that of HZO1/1.

4. Discussion

Three capacitors were investigated: ZrO_2/HfO_2 = 1/1 (HZO1/1) was the most common deposition ratio, ZrO_2/HfO_2 = 5/5 (HZO5/5) was the fully laminated structure, and the ZrO_2/HfO_2 = 5/5 laminated structure as an interlayer (HZO1/5/1) was proposed in this work. The electrical test results showed that the remanent polarization was significantly reduced, but the endurance was greatly improved by two orders of magnitude in the fully laminated structure HZO5/5. The advantages of the high endurance of HZO5/5 and high remanent polarization of HZO1/1 were combined in HZO1/5/1. In HZO1/5/1, the endurance was also improved by two orders of magnitude without a significant reduction in the remanent polarization. The internal mechanism of the laminated structure was analyzed in terms of the remanent polarization and endurance performance.

The remanent polarization was an important factor to measure the ferroelectricity of ferroelectric devices. The experimental results showed that the ferroelectric properties of the thin films were affected by changing the ZrO_2/HfO_2 deposition ratio. During the crystallization process, the formation of the ferroelectric phase was thought to be the result of inhibiting the tetragonal (T) phase to monoclinic (M) phase transformation, which leads to the formation of the asymmetric orthorhombic (O) phase [1]. Previous studies reported that $Hf_{0.5}Zr_{0.5}O_2$ exhibits optimal ferroelectricity [37]. In the laminated structure, only part of HfO_2 and ZrO_2 are completely miscible at the ZrO_2/HfO_2 interfaces, and $Hf_{0.5}Zr_{0.5}O_2$ is formed. The incompletely miscible parts of HfO_2 and ZrO_2 are shown as less Zr doping in HfO_2 and more Zr doping in ZrO_2, respectively. More of the M-phase fraction is shown in less Zr-doped HfO_2 films, and more of the T-phase fraction is shown in rich Zr-doped HfO_2 films [17]. The O-phase fraction decreased because of the increase in the T phase or M phase in incompletely miscible parts, which led to the reduction in the remanent polarization of the laminated structure. Therefore, the influence of remanent polarization could be effectively reduced in HZO1/5/1 compared with the fully laminated structure HZO5/5 because the thickness of the laminated structure was small, only 1/6 of the thickness of the whole film.

The leakage current and the breakdown behavior were attributed to the migration and aggregation of defects. Oxygen vacancy is generally regarded as the dominant defect in ferroelectric capacitors. It is reported that the migration barriers from interfaces in HZO and laminated structures are different [38]. The vertical migration of oxygen vacancies can be suppressed in a laminated structure due to the higher energy barriers. For HZO5/5 and HZO1/5/1, since the migration of oxygen vacancies was suppressed in the laminated-structure, the increase in the leakage current was suppressed, resulting in an improved endurance.

The improved reliability may also be attributed to the nonuniform distribution of defects caused by the laminated-structure interlayer, as shown in Figure 7a,b. The vertical migration and accumulation distribution of the oxygen vacancies promoted the generation and development of conduction in the current path, as shown in Figure 7a. In the HfO_2 and ZrO_2 layers, the distributions of the oxygen vacancies were different [39]. The formation of highly conductive filaments in HfO_2 was accompanied by phase decomposition into hexagonal metal phases, such as Hf and h-Hf_6O [40]. Compared with ZrO_x, HfO_x had a strong concentration force for oxygen vacancies to produce highly conductive filaments. Under annealing, oxygen vacancies were oriented towards the TiN electrode side, which merged to form conductive filaments where the tip moved towards another TiN electrode side. As shown in Figure 7b, oxygen vacancies tended to accumulate strongly in HfO_2, but diffusively in ZrO_2 without obvious accumulation [40]. The non-aggregation of oxygen vacancies in thicker ZrO_2 layers was difficult for forming a conductive path.

Therefore, lower leakage current and higher reliability were achieved in the capacitor with the laminated-structure interlayer.

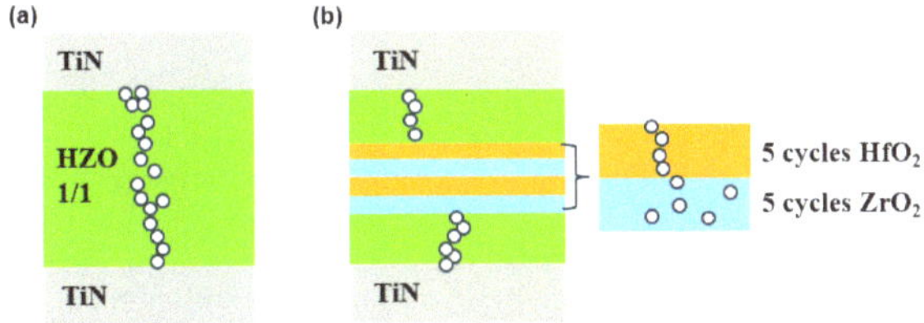

Figure 7. Schematic diagrams of the physical mechanism of the breakdown behavior of (**a**) HZO1/1 and (**b**) HZO1/5/1 devices.

It is worth noting that the improved endurance of the laminated-structure interlayer was not without negative impacts. The crystallization of the ferroelectric film and the fraction of the phases may have affected the ferroelectric properties of the capacitor. Therefore, it is necessary to carefully balance the improved endurance and the potential negative impact on the ferroelectric properties when using the laminated-structure interlayer. Overall, the analysis of the remanent polarization and endurance provides a deeper understanding of the internal mechanism behind the improved reliability of ferroelectric capacitors with a laminated-structure interlayer.

5. Conclusions

In this article, the present study demonstrated the feasibility of using a laminated-structure interlayer as a simple and effective method to improve the endurance of $Hf_{0.5}Zr_{0.5}O_2$ ferroelectric thin films without additional materials and a complicated process. The laminated-structure interlayer was deposited by altering the ZrO_2–HfO_2 deposition ratio from 1:1 to 5:5. From the electrical tests and physical characterization, the optimized capacitor exhibited one order of magnitude of reduction in the leakage current and excellent endurance performance improved by two orders of magnitude (from 10^6 to 10^8 cycles) compared with HZO1/1. Moreover, the Time-Dependent Dielectric Breakdown (TDDB) reliability was enhanced and the breakdown voltage was increased. These results are ascribed to a different phase fraction, migration inhibition and the nonuniform distribution of oxygen vacancies in the $Hf_{0.5}Zr_{0.5}O_2$ thin film with a laminated-structure interlayer. This work provides a feasible strategy by a deposition ratio adjustment, which contributes to enhancing the reliability of $Hf_{0.5}Zr_{0.5}O_2$ ferroelectric films in nonvolatile memory devices.

Author Contributions: Conceptualization, Y.W. (Yan Wang) and M.C.; methodology, M.C. and S.L.; investigation, M.C., B.W., P.J. and Y.D.; data curation, Y.W. (Yuan Wang), M.C. and Y.C. (Yuting Chen); writing—original draft preparation, M.C.; formal analysis, M.C. and Y.C. (Yuanxiang Chen); writing—review and editing, Y.W. (Yan Wang); supervision, Y.W. (Yan Wang); project administration, Y.W. (Yan Wang); funding acquisition, Y.W. (Yan Wang). All authors have read and agreed to the published version of the manuscript.

Funding: This work was supported in part by the Ministry of Science and Technology (MOST) of China under Grant 2022YFB3608400; in part by the Strategic Priority Research Program of the Chinese Academy of Sciences under Grant XDB44000000, in part by the National Natural Science Foundation of China under Grant U21B2030.

Data Availability Statement: The data presented in this study are available on request from the corresponding author.

Conflicts of Interest: The authors declare no conflict of interest.

References

1. Böscke, T.S.; Müller, J.; Bräuhaus, D.; Schröder, U.; Böttger, U. Ferroelectricity in hafnium oxide thin films. *Appl. Phys. Lett.* **2011**, *99*, 102903. [CrossRef]
2. Park, M.H.; Lee, Y.H.; Mikolajick, T.; Schroeder, U.; Hwang, C.S. Review and perspective on ferroelectric HfO_2-based thin films for memory applications. *MRS Commun.* **2018**, *8*, 795–808. [CrossRef]
3. Chouprik, A.; Chernikova, A.; Markeev, A.; Mikheev, V.; Negrov, D.; Spiridonov, M.; Zarubin, S.; Zenkevich, A. Electron transport across ultrathin ferroelectric $Hf_{0.5}Zr_{0.5}O_2$ films on Si. *Microelectron. Eng.* **2017**, *178*, 250–253. [CrossRef]
4. Tian, X.; Shibayama, S.; Nishimura, T.; Yajima, T.; Migita, S.; Toriumi, A. Evolution of ferroelectric HfO_2 in ultrathin region down to 3 nm. *Appl. Phys. Lett.* **2018**, *112*, 102902. [CrossRef]
5. Chen, L.; Wang, T.Y.; Dai, Y.W.; Cha, M.Y.; Zhu, H.; Sun, Q.Q.; Ding, S.J.; Zhou, P.; Chua, L.; Zhang, D.W. Ultra-low power $Hf_{0.5}Zr_{0.5}O_2$ based ferroelectric tunnel junction synapses for hardware neural network applications. *Nanoscale* **2018**, *10*, 15826–15833. [CrossRef] [PubMed]
6. Siannas, N.; Zacharaki, C.; Tsipas, P.; Chaitoglou, S.; Begon-Lours, L.; Dimoulas, A. Scaling ferroelectric HZO thickness for low power Ge MFS-FTJ memories. In Proceedings of the ESSDERC 2021-IEEE 51st European Solid-State Device Research Conference (ESSDERC), Grenoble, France, 13–22 September 2021; pp. 287–290.
7. Hoffmann, M.; Tan, A.J.; Shanker, N.; Liao, Y.-H.; Wang, L.-C.; Bae, J.-H.; Hu, C.; Salahuddin, S. Fast Read-After-Write and Depolarization Fields in High Endurance n-Type Ferroelectric FETs. *IEEE Electron Device Lett.* **2022**, *43*, 717–720. [CrossRef]
8. Müller, J.; Böscke, T.; Müller, S.; Yurchuk, E.; Polakowski, P.; Paul, J.; Martin, D.; Schenk, T.; Khullar, K.; Kersch, A. Ferroelectric hafnium oxide: A CMOS-compatible and highly scalable approach to future ferroelectric memories. In Proceedings of the 2013 IEEE International Electron Devices Meeting, Washington, DC, USA, 9–11 December 2013; pp. 10.8.1–10.8.4.
9. Sang, X.; Grimley, E.D.; Schenk, T.; Schroeder, U.; LeBeau, J.M. On the structural origins of ferroelectricity in HfO_2 thin films. *Appl. Phys. Lett.* **2015**, *106*, 162905. [CrossRef]
10. Zhao, C.; Roebben, G.; Heyns, M.; Van der Biest, O. Crystallisation and tetragonal-monoclinic transformation in ZrO_2 and HfO_2 dielectric thin films. *Key Eng. Mater.* **2002**, *206*, 1285–1288.
11. Müller, J.; Schröder, U.; Böscke, T.S.; Müller, I.; Böttger, U.; Wilde, L.; Sundqvist, J.; Lemberger, M.; Kücher, P.; Mikolajick, T.; et al. Ferroelectricity in yttrium-doped hafnium oxide. *J. Appl. Phys.* **2011**, *110*, 114113. [CrossRef]
12. Müller, J.; Böscke, T.S.; Bräuhaus, D.; Schröder, U.; Böttger, U.; Sundqvist, J.; Kücher, P.; Mikolajick, T.; Frey, L. Ferroelectric $Zr_{0.5}Hf_{0.5}O_2$ thin films for nonvolatile memory applications. *Appl. Phys. Lett.* **2011**, *99*, 112901. [CrossRef]
13. Chernikova, A.G.; Kuzmichev, D.S.; Negrov, D.V.; Kozodaev, M.G.; Polyakov, S.N.; Markeev, A.M. Ferroelectric properties of full plasma-enhanced ALD TiN/La:HfO2/TiN stacks. *Appl. Phys. Lett.* **2016**, *108*, 242905. [CrossRef]
14. Mueller, S.; Mueller, J.; Singh, A.; Riedel, S.; Sundqvist, J.; Schroeder, U.; Mikolajick, T. Incipient Ferroelectricity in Al-Doped HfO_2 Thin Films. *Adv. Funct. Mater.* **2012**, *22*, 2412–2417. [CrossRef]
15. Schenk, T.; Mueller, S.; Schroeder, U.; Materlik, R.; Kersch, A.; Popovici, M.; Adelmann, C.; Van Elshocht, S.; Mikolajick, T. Strontium doped hafnium oxide thin films: Wide process window for ferroelectric memories. In Proceedings of the 2013 European Solid-State Device Research Conference (ESSDERC), Bucharest, Romania, 16–20 September 2013; pp. 260–263.
16. Kim, S.K.; Hwang, C.S.J.E.; Letters, S.-S. Atomic layer deposition of ZrO_2 thin films with high dielectric constant on TiN substrates. *Electrochem. Solid-State Lett.* **2007**, *11*, G9. [CrossRef]
17. Muller, J.; Boscke, T.S.; Schroder, U.; Mueller, S.; Brauhaus, D.; Bottger, U.; Frey, L.; Mikolajick, T. Ferroelectricity in Simple Binary ZrO_2 and HfO_2. *Nano. Lett.* **2012**, *12*, 4318–4323. [CrossRef] [PubMed]
18. Kim, H.J.; Park, M.H.; Kim, Y.J.; Lee, Y.H.; Moon, T.; Kim, K.D.; Hyun, S.D.; Hwang, C.S. A study on the wake-up effect of ferroelectric $Hf_{0.5}Zr_{0.5}O_2$ films by pulse-switching measurement. *Nanoscale* **2016**, *8*, 1383–1389. [CrossRef]
19. Chernikova, A.G.; Kozodaev, M.G.; Negrov, D.V.; Korostylev, E.V.; Park, M.H.; Schroeder, U.; Hwang, C.S.; Markeev, A.M. Improved Ferroelectric Switching Endurance of La-Doped $Hf_{0.5}Zr_{0.5}O_2$ Thin Films. *ACS Appl. Mater. Interfaces* **2018**, *10*, 2701–2708. [CrossRef]
20. Jiang, P.; Yang, Y.; Wei, W.; Gong, T.; Wang, Y.; Chen, Y.; Ding, Y.; Lv, S.; Wang, B.; Chen, M. Stress Effects of Interconnecting Metals on Back-End-of-Line Compatible $Hf_{0.5}Zr_{0.5}O_2$ Ferroelectric Capacitors. *IEEE Electron Device Lett.* **2023**, *44*, 602–605. [CrossRef]
21. Cao, R.; Song, B.; Shang, D.; Yang, Y.; Luo, Q.; Wu, S.; Li, Y.; Wang, Y.; Lv, H.; Liu, Q. Improvement of endurance in HZO-based ferroelectric capacitor using Ru electrode. *IEEE Electron Device Lett.* **2019**, *40*, 1744–1747. [CrossRef]
22. Walke, A.M.; Popovici, M.I.; Banerjee, K.; Clima, S.; Kumbhare, P.; Desmet, J.; Meersschaut, J.; Van den Bosch, G.; Delhougne, R.; Kar, G.S. Electrical Investigation of Wake-Up in High Endurance Fatigue-Free La and Y Doped HZO Metal–Ferroelectric–Metal Capacitors. *IEEE Trans. Electron Devices* **2022**, *69*, 4744–4749. [CrossRef]
23. Choi, Y.; Han, C.; Shin, J.; Moon, S.; Min, J.; Park, H.; Eom, D.; Lee, J.; Shin, C. Impact of Chamber/Annealing Temperature on the Endurance Characteristic of Zr:HfO_2 Ferroelectric Capacitor. *Sensors* **2022**, *22*, 4087. [CrossRef]
24. Xu, Y.; Yang, Y.; Zhao, S.; Gong, T.; Jiang, P.; Lv, S.; Yu, H.; Yuan, P.; Dang, Z.; Ding, Y.; et al. Robust Breakdown Reliability and Improved Endurance in $Hf_{0.5}Zr_{0.5}O_2$ Ferroelectric Using Grain Boundary Interruption. *IEEE Trans. Electron Devices* **2022**, *69*, 430–433. [CrossRef]
25. Lu, Y.; Shieh, J.; Tsai, F. Induction of ferroelectricity in nanoscale ZrO_2/HfO_2 bilayer thin films on $Pt/Ti/SiO_2/Si$ substrates. *Acta Mater.* **2016**, *115*, 68–75. [CrossRef]

26. Weeks, S.L.; Pal, A.; Narasimhan, V.K.; Littau, K.A.; Chiang, T. Engineering of ferroelectric HfO_2-ZrO_2 nanolaminates. *ACS Appl. Mater. Interfaces* **2017**, *9*, 13440–13447. [CrossRef] [PubMed]
27. Liang, Y.-K.; Li, W.-L.; Wang, Y.-J.; Peng, L.-C.; Lu, C.-C.; Huang, H.-Y.; Yeong, S.H.; Lin, Y.-M.; Chu, Y.-H.; Chang, E.-Y. ZrO_2-HfO_2 Superlattice Ferroelectric Capacitors with Optimized Annealing to Achieve Extremely High Polarization Stability. *IEEE Electron Device Lett.* **2022**, *43*, 1451–1454. [CrossRef]
28. Lehninger, D.; Prabhu, A.; Sünbül, A.; Ali, T.; Schöne, F.; Kämpfe, T.; Biedermann, K.; Roy, L.; Seidel, K.; Lederer, M. Ferroelectric [HfO_2/ZrO_2] Superlattices with Enhanced Polarization, Tailored Coercive Field, and Improved High Temperature Reliability. *Adv. Phys. Res.* **2023**. [CrossRef]
29. Peng, Y.; Xiao, W.; Liu, Y.; Jin, C.; Deng, X.; Zhang, Y.; Liu, F.; Zheng, Y.; Cheng, Y.; Chen, B. HfO_2-ZrO_2 superlattice ferroelectric capacitor with improved endurance performance and higher fatigue recovery capability. *IEEE Electron Device Lett.* **2021**, *43*, 216–219. [CrossRef]
30. Park, M.H.; Kim, H.J.; Lee, G.; Park, J.; Lee, Y.H.; Kim, Y.J.; Moon, T.; Kim, K.D.; Hyun, S.D.; Park, H.W. A comprehensive study on the mechanism of ferroelectric phase formation in hafnia-zirconia nanolaminates and superlattices. *Appl. Phys. Rev.* **2019**, *6*, 041403. [CrossRef]
31. Jiang, P.; Wei, W.; Yang, Y.; Wang, Y.; Xu, Y.; Tai, L.; Yuan, P.; Chen, Y.; Gao, Z.; Gong, T. Stabilizing Remanent Polarization during Cycling in HZO-Based Ferroelectric Device by Prolonging Wake-up Period. *Adv. Electron. Mater.* **2022**, *8*, 2100662. [CrossRef]
32. Park, M.H.; Lee, Y.H.; Kim, H.J.; Kim, Y.J.; Moon, T.; Kim, K.D.; Hyun, S.D.; Hwang, C.S. Morphotropic phase boundary of $Hf_{1-x}Zr_xO_2$ thin films for dynamic random access memories. *ACS Appl. Mater. Interfaces* **2018**, *10*, 42666–42673. [CrossRef]
33. Sun, Z.; Ma, C.; Liu, M.; Cui, J.; Lu, L.; Lu, J.; Lou, X.; Jin, L.; Wang, H.; Jia, C.L. Ultrahigh energy storage performance of lead-free oxide multilayer film capacitors via interface engineering. *Adv. Mater.* **2017**, *29*, 1604427. [CrossRef]
34. Chen, Y.; Wang, L.; Liu, L.; Tang, L.; Yuan, X.; Chen, H.; Zhou, K.; Zhang, D. Thickness-dependent ferroelectric properties of HfO_2/ZrO_2 nanolaminates using atomic layer deposition. *J. Mater. Sci.* **2021**, *56*, 6064–6072. [CrossRef]
35. Degraeve, R.; Groeseneken, G.; Bellens, R.; Ogier, J.L.; Depas, M.; Roussel, P.J.; Maes, H.E. New insights in the relation between electron trap generation and the statistical properties of oxide breakdown. *IEEE Trans. Electron Devices* **1998**, *45*, 904–911. [CrossRef]
36. Chen, Y.-H.; Su, C.-J.; Yang, T.-H.; Hu, C.; Wu, T.-L. Improved TDDB Reliability and Interface States in 5-nm $Hf_{0.5}Zr_{0.5}O_2$ Ferroelectric Technologies Using NH_3 Plasma and Microwave Annealing. *IEEE Trans. Electron Devices* **2020**, *67*, 1581–1585. [CrossRef]
37. Kim, S.J.; Mohan, J.; Summerfelt, S.R.; Kim, J. Ferroelectric $Hf_{0.5}Zr_{0.5}O_2$ thin films: A review of recent advances. *JOM* **2019**, *71*, 246–255. [CrossRef]
38. Gong, Z.; Chen, J.; Peng, Y.; Liu, Y.; Yu, X.; Han, G. Physical origin of the endurance improvement for HfO_2-ZrO_2 superlattice ferroelectric film. *Appl. Phys. Lett.* **2022**, *121*, 242901. [CrossRef]
39. Bai, N.; Xue, K.H.; Huang, J.; Yuan, J.H.; Wang, W.; Mao, G.Q.; Zou, L.; Yang, S.; Lu, H.; Sun, H.; et al. Designing Wake-Up Free Ferroelectric Capacitors Based on the HfO. Superlattice Structure. *Adv. Electron. Mater.* **2022**, *9*, 2200737. [CrossRef]
40. Xue, K.-H.; Miao, X.-S. Oxygen vacancy chain and conductive filament formation in hafnia. *J. Appl. Phys.* **2018**, *123*, 161505. [CrossRef]

nanomaterials

MDPI

Communication

Comprehensive Investigation of Constant Voltage Stress Time-Dependent Breakdown and Cycle-to-Breakdown Reliability in Y-Doped and Si-Doped HfO_2 Metal-Ferroelectric-Metal Memory

Ting-Yu Chang [1], Kuan-Chi Wang [1], Hsien-Yang Liu [2], Jing-Hua Hseun [3], Wei-Cheng Peng [1], Nicolò Ronchi [4], Umberto Celano [4,5], Kaustuv Banerjee [4], Jan Van Houdt [4,6] and Tian-Li Wu [1,2,3,*]

1 International College of Semiconductor Technology, National Yang Ming Chiao Tung University, Hsinchu 30010, Taiwan; emmatony11111.st10@nycu.edu.tw (T.-Y.C.); kcwang.st09@nycu.edu.tw (K.-C.W.); wcpeng.st11@nycu.edu.tw (W.-C.P.)
2 Institute of Electronics, National Yang Ming Chiao Tung University, Hsinchu 30010, Taiwan; hyliu.ee10@nycu.edu.tw
3 Institute of Pioneer Semiconductor Innovation, National Yang Ming Chiao Tung University, Hsinchu 30010, Taiwan; jhhsuen.10@nycu.edu.tw
4 Imec, 3000 Leuven, Belgium; nicolo.ronchi@imec.be (N.R.); umberto.celano@asu.edu (U.C.); kaustuv.banerjee@imec.be (K.B.); jan.vanhoudt@imec.be (J.V.H.)
5 School of Electrical, Computer and Energy Engineering, Arizona State University, Tempe, AZ 85287, USA
6 Department of Physics and Astronomy, KU Leuven, 3000 Leuven, Belgium
* Correspondence: tlwu@nycu.edu.tw

Abstract: In this study, we comprehensively investigate the constant voltage stress (CVS) time-dependent breakdown and cycle-to-breakdown while considering metal-ferroelectric-metal (MFM) memory, which has distinct domain sizes induced by different doping species, i.e., Yttrium (Y) (Sample A) and Silicon (Si) (Sample B). Firstly, Y-doped and Si-doped HfO_2 MFM devices exhibit domain sizes of 5.64 nm and 12.47 nm, respectively. Secondly, Si-doped HfO_2 MFM devices (Sample B) have better CVS time-dependent breakdown and cycle-to-breakdown stability than Y-doped HfO_2 MFM devices (Sample A). Therefore, a larger domain size showing higher extrapolated voltage under CVS time-dependent breakdown and cycle-to-breakdown evaluations was observed, indicating that the domain size crucially impacts the stability of MFM memory.

Keywords: ferroelectric; domain size; reliability

1. Introduction

Since the initial discovery of Si-doped HfO_2 materials' ferroelectric properties in 2011 [1], significant attention has been directed toward oxide materials with a fluorite structure, such as doped HfO_2 [2], and the solid solution of $Hf_xZr_{1-x}O$ (HZO) [3]. These materials have garnered interest for their compatibility with advanced process technology and ability to exhibit ferroelectric behavior even at thicknesses of $\leq$10 nm, setting them apart from traditional perovskite ferroelectric materials. Recent studies have revealed that the thickness of HfO_2-based ferroelectric films can be reduced to as little as 1 nm while maintaining the occurrence of spontaneous polarization and its ability to alter polarization direction [4]. This finding suggests that HfO_2-based ferroelectric film does not have a critical threshold for scaling down, unlike perovskite materials. This exceptional scalability feature indicates a promising advantage for developing memory devices driven by polarization.

Furthermore, ferroelectric HfO_2-based technologies are promising materials for non-volatile memories [5], logic FETs [6], and neuromorphic applications [7–9] because of their compatibility with complementary metal-oxide-semiconductor (CMOS) technology [10,11]. Ferroelectric properties can be induced by various doping species in HfO_2 films, e.g., Zr, Si, Al, Gd, etc. Recently, high-performance ferroelectric-based technologies have been demonstrated with optimized annealing conditions, dopants, electrodes, interfacial layers, etc. [12–21]. However, reliability remains one of the main concerns in ferroelectric-based technologies [22], particularly the instability related to domain size's impact on time-dependent dielectric breakdown and cycle-to-breakdown. Understanding the impact of domain size on the stability of ferroelectric-based devices is not extensively reported in the literature.

In this study, Yttrium (Y)-doped and Silicon (Si)-doped HfO_2 metal-ferroelectric-metal (MFM) devices were fabricated to intentionally induce different domain sizes in metal-ferroelectric-metal (MFM) devices. The o-phase with ferroelectricity can be induced through the annealing process of differently doped ferroelectric films since the crystal radius of doping below/above Hf can stabilize the t-/c-phases [2]. CVS time-dependent stress and cycle-to-breakdown measurements were conducted. Furthermore, the correlations between domain size and CVS time-dependent stress and cycle-to-breakdown stability are discussed and analyzed to understand the impact of domain size.

2. Materials and Methods

Figure 1 shows the schematic structure of metal-ferroelectric-metal (MFM) capacitors and a brief process flow of this work. At first, 10-nm TiN was deposited via PVD as the bottom electrode. Next, 9.5-nm HfO_2-based ferroelectric layers with two different dopants, Y and Si, were deposited via thermal ALD at 300 °C. Afterward, another 10-nm TiN was deposited via ALD on the ferroelectric films as the top metal electrode. Lastly, RTA was conducted in N_2 ambient at 650 °C for crystallization for 20 s.

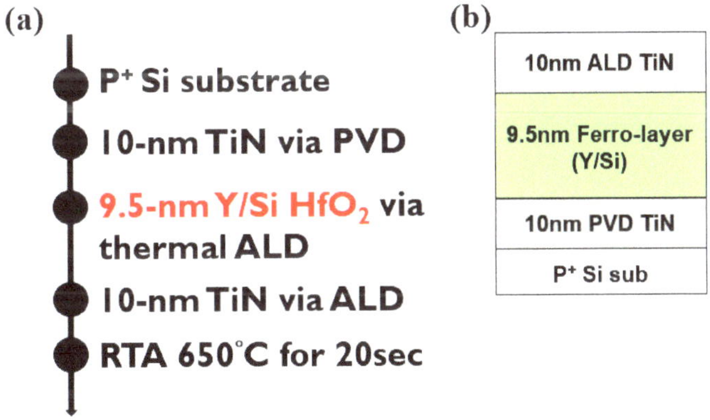

Figure 1. (**a**) Schematic of the process flow and (**b**) schematic structure of MFM capacitors with different dopants in ferroelectric layers.

The measurement setup for the capacitor used in this study was 2400 μm^2 (60 μm × 80 μm). To perform electrical characterizations such as I-V (current-voltage) and time-dependent dielectric breakdown (TDDB) measurements, a Keysight B1500 Source Measurement Unit (SMU) Keysight, USA, was employed. To characterize the ferroelectric properties, including P-V (polarization-voltage) and cycle-to-breakdown measurements, a Keysight B1530 Waveform Generator/Function Measurement Unit (WGFMU) Keysight, USA, was utilized. In this setup, the capacitor was biased at the bottom via a chuck electrode, while a ground electrode was placed on top.

To compare the sample's domain size, distribution, and homogeneity, we used contact resonance piezoresponse force microscopy (PFM). Although a quantitative interpretation of the results is beyond the scope of this work, we used the same probe. We also operated under the same conditions on two samples with the same physical thickness (9.5 nm). Therefore, the results represent a relative comparison between samples and can be used to analyze the domain size (Figure 2). For instance, Sample A (Y-doped) has a smaller domain size than Sample B (Si-doped). Details of the domain structures can also be found elsewhere [23]. Table 1 briefly describes the dopants and domain sizes used in this study. Distinct differences in the domain sizes of Sample A and Sample B can be used to understand the impact of domain size on the reliability of MFM devices.

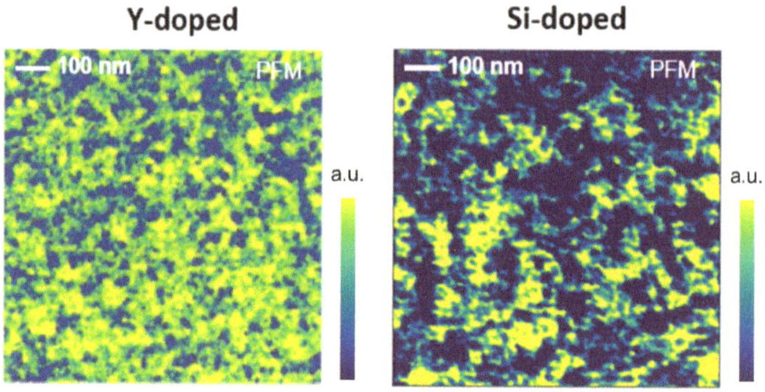

Figure 2. Contact resonance piezoresponse force microscopy PFM measurements in Sample A (Y-doped) and Sample B (Si-doped).

Table 1. A brief summary of the dopants and domain sizes used in this study.

	Dopant in Ferroelectric Layer	Domain Size (nm)	Doping Concentration
Sample A	Y	5.6	2.9%
Sample B	Si	12.5	2.5%

3. Results

To understand ferroelectricity, we used P-V measurements with a triangular pulse of 10 μs/V and a trapezoidal plus with T_r (rising time)/T_f (falling time) fixed at 0.5 μs and T_{width} (pulse width) set at 1 μs for the cycling. Figure 3 shows the P-V characteristics of Samples A and B in the fresh state and after 10^3 cycles. Figure 4 shows comprehensive endurance characteristics at different cycling numbers. Sample A shows a slightly larger 2Pr than Sample B in the pristine state. Upon increasing the cycling number, Sample A exhibits a clear wake-up effect with a saturation of 2Pr after 10^4 cycles. However, Sample B does not exhibit a saturation of 2Pr. Overall, Sample A shows a larger 2Pr than Sample B after 10^5 cycles

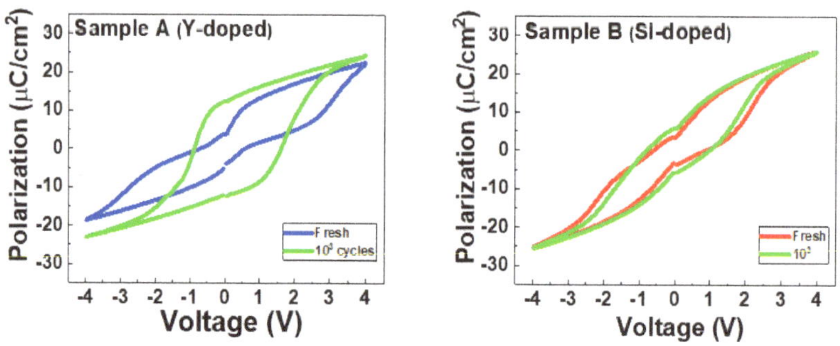

Figure 3. P-V characteristics in Samples A and B.

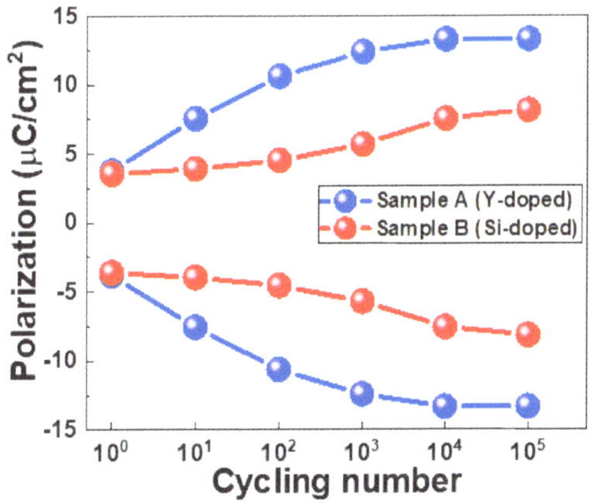

Figure 4. Polarization characteristics with respect to cycling numbers.

To understand the impact of domain size on time-dependent breakdown stability, we performed constant voltage stress time-dependent dielectric breakdown and cycling-to-breakdown evaluations. Figures 5 and 6 show the results of the constant voltage stress (CVS) time-dependent dielectric breakdown (TDDB) and cycle-to-breakdown evaluations in Samples A and B, respectively.

Figure 5c,d shows Weibull plots of time-to-breakdown (t_{BD}) distributions for three TDDB VG conditions, which follow the Weibull failure distribution:

$$\ln[-\ln(1-\mathrm{F(t)})] = \beta \ln(t) - \beta \ln(\eta) \tag{1}$$

where t is the time; β is the shape parameter; η is the scale factor of 63.2% value. The fitted β is 2.497 and 3.829 for Samples A and B, respectively. A higher β implies a tight distribution and small variability.

In Figure 5e, lifetime of 1% failure analyses are extrapolated from the Weibull plot of t_{BD} distribution and projected to a 10-year line. The operating voltages of Sample A are slightly lower than Sample B (2.62 V) at 2.24 V.

In designing the cycle-to-breakdown (Cycle-to-BD) measurement, we chose a PUND waveform with triangular pulses for the reading state. We set the rising, falling, and delay times ($T_r/T_f/T_{delay}$) to a fixed duration of 5 μs. The PUND waveform allowed for clear observation of whether or not the sample experienced breakdown. In the cycling state, trapezoidal pulses were used as the waveform. The T_r and T_f were fixed at 0.5 μs, while the T_{width} was set to 1 μs. To determine the cycles of the chosen reading step, we divided the interval to reach 1E6 cycles into 12 segments. After calculations, we determined that $10^{0.375}$ cycles would serve as the interval between the two consecutive reading states.

Similar to the TDDB analysis, cycle-to-breakdown (Cycle-to-BD) distributions for three different Cycle-to-BD VG conditions were used to construct Weibull plots. These plots were then used to extract fitting values from the ß value and generate lifetime curves. However, to adapt the Weibull failure distribution to the Cycle-to-BD analysis, the time-to-BD (t_{BD}) was transformed into cycle-to-BD (C_{BD}). Figure 6c,d display the fitting β values and corresponding lines for different devices. Sample A yielded a fitted β value of 0.948, while sample B had a βvalue of 1.274. These values are consistent with the TDDB analysis results, where Sample B exhibited a higher β value. The discrepancy in β values between TDDB and Cycle-to-BD measurements may be due to the Cycle-to-BD measurement using both positive and negative pulses during the cycling stage compared to TDDB measurements with a constant positive bias.

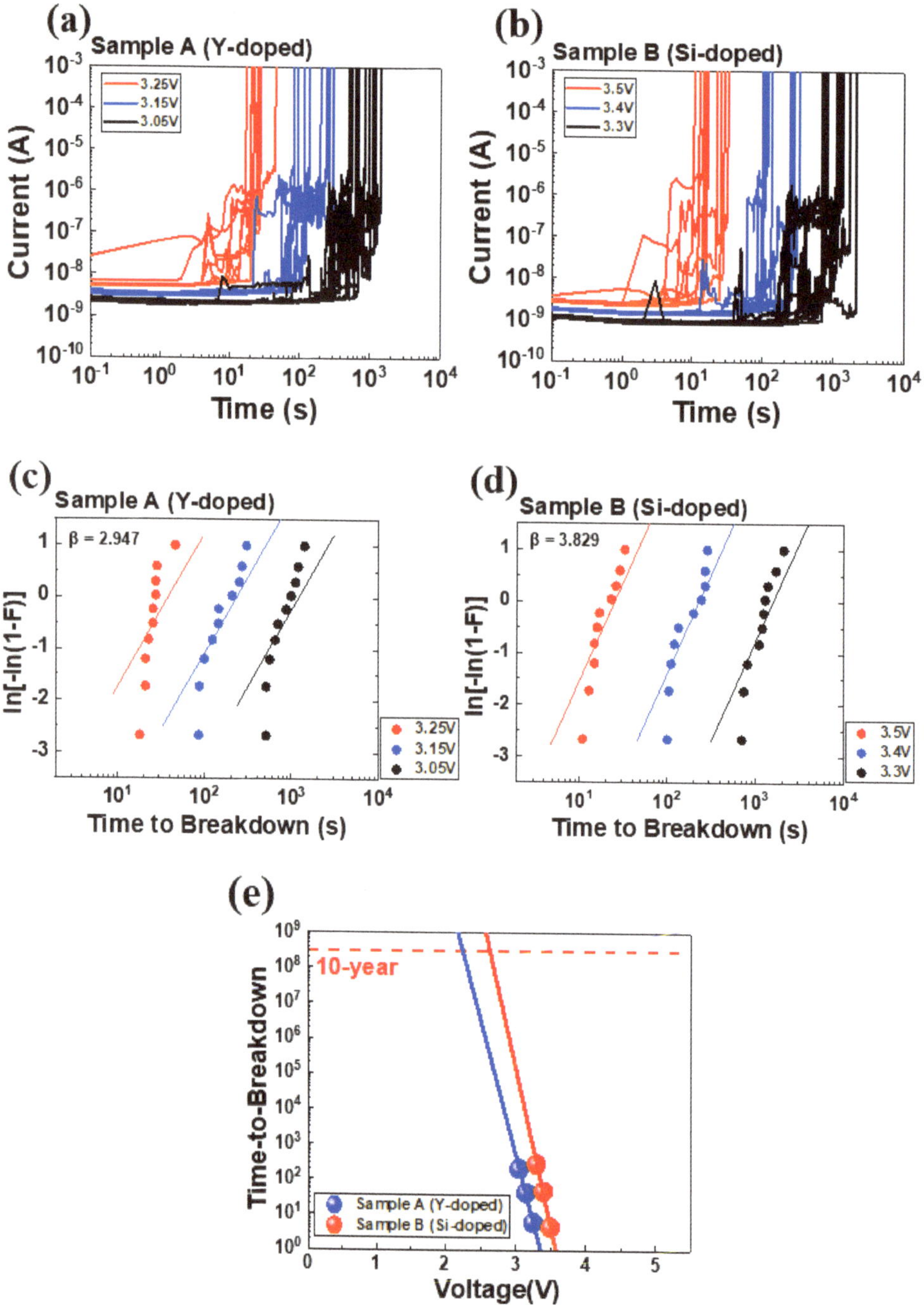

Figure 5. Gate current for (**a**) Sample A and (**b**) Sample B, as monitored by six different voltages (10 devices per group). (**c**,**d**) show corresponding Weibull plots of t_{BD}. (**e**) Operating voltage extrapolation for a 10-year lifetime at 1% failure for devices with Samples A and B.

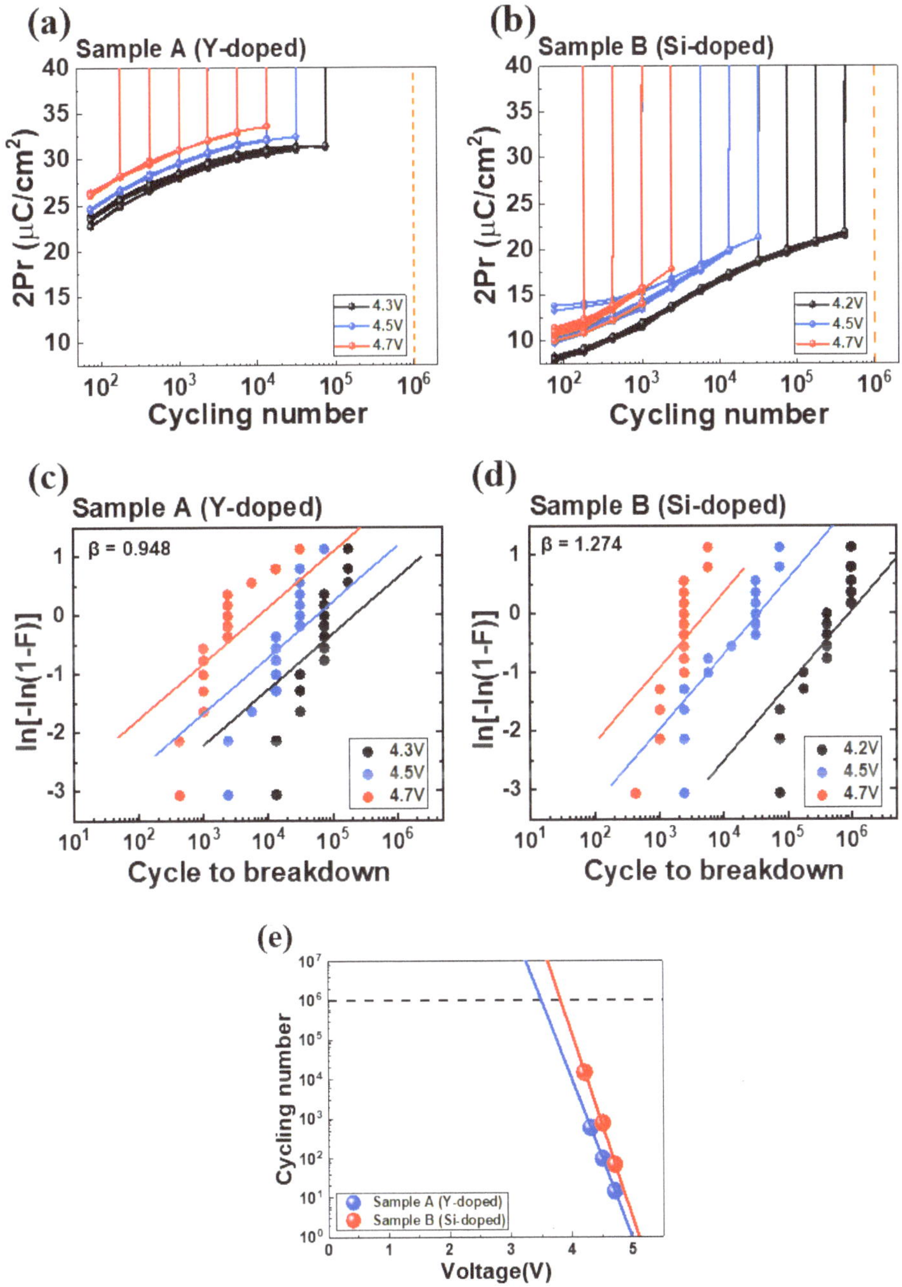

Figure 6. 2Pr for (**a**) Sample A and (**b**) Sample B, as monitored by six different voltages (10 devices per group). (**c**,**d**) show corresponding Weibull plots of cycle-to-breakdown. (**e**) Operating cycling number extrapolation for a 10^6-cycling lifetime at 1% failure for Samples A and B.

Figure 6e illustrates the 1% failure analysis of the 10^6-cycle lifetime analysis. The trend observed in the Cycle-to-BD analysis is similar to the TDDB analysis. For Sample A, the determined operating voltage is 3.45 V, which is slightly lower than Sample B's operating voltage of 3.79 V.

Figure 7 shows the correlation between operation voltage (based on constant voltage stress TDDB and cycle-to-breakdown measurements) and domain size. Table 2 summarizes operation voltages based on CVS TDDB and cycle-to-breakdown evaluations. Figure 7 indicates that a larger domain size exhibits better TDDB and cycle-to-breakdown stability, i.e., higher 10-year operation voltages and higher voltages up to 10^6 cycles. In addition, Table 3 presents the maximum lifetime of targeting applied voltage at 3V results that Samples A and B can withstand in seconds. We calculated the lifetime of the Cycle-to-BD measurement by multiplying the predicted breakdown cycling number by a duration of 1 cycle in the measurement waveform design. According to our results, the lifetime measured in the TDDB analysis is longer than in the Cycle-to-BD analysis. This finding indicates that the measurement technique involving the continuous application of the same bias direction is less likely to induce defects generation and device breakdown than the technique involving a continuously changing bias direction. However, from the lifetimes in seconds, Sample B still exhibits a longer lifetime with a larger domain size in both the TDDB and Cycle-to-BD measurements. These findings are consistent with previous results.

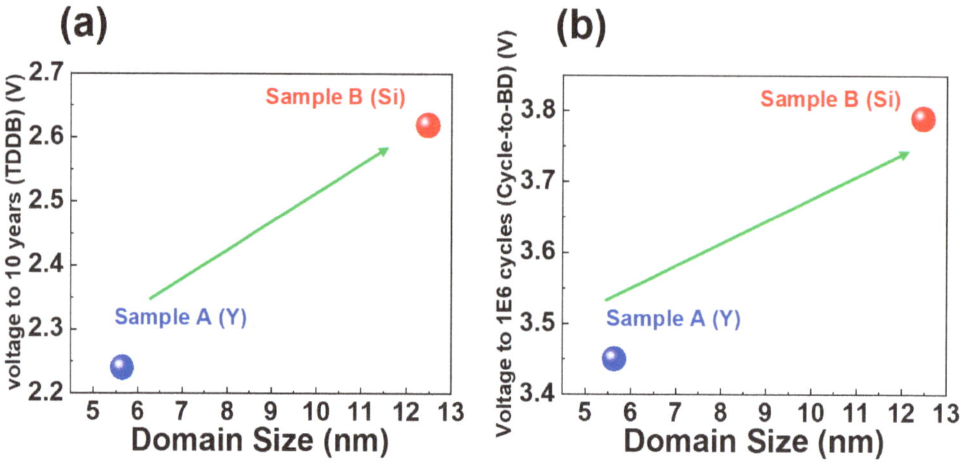

Figure 7. Extrapolated voltages vs. domain size under CVS TDDB (**a**) and cycle-to-breakdown measurements (**b**).

Table 2. Summary of the extrapolated operation voltage.

	10-year Operation Voltage Based on CVS TDDB	10^6-Cycling Operation Voltage Based on Cycle-to-Breakdown
Sample A	2.24	3.45
Sample B	2.62	3.79

Table 3. The maximum lifetime results that two samples can withstand in seconds.

	Lifetime of Cycle-to-BD @3V (s)	Lifetime of TDDB @3V (s)
Sample A	203	515
Sample B	12,170	146,058

The charging effects along the domain boundaries have been reported [24]. Therefore, larger domain sizes represent fewer domain boundaries (Figure 8) and reduce the chance of charging effects that create leakage paths as bias is applied to the device, thereby improving TDDB and cycle-to-breakdown stability. Besides, it is worth noting that Sample B exhibits larger β than Sample A, indicating better uniformity due to fewer domain boundaries in Sample B. In summary, larger domain sizes and fewer domain boundaries can improve CVS TDDB and cycle-to-breakdown reliability. However, samples with a larger domain size may degrade ferroelectricity.

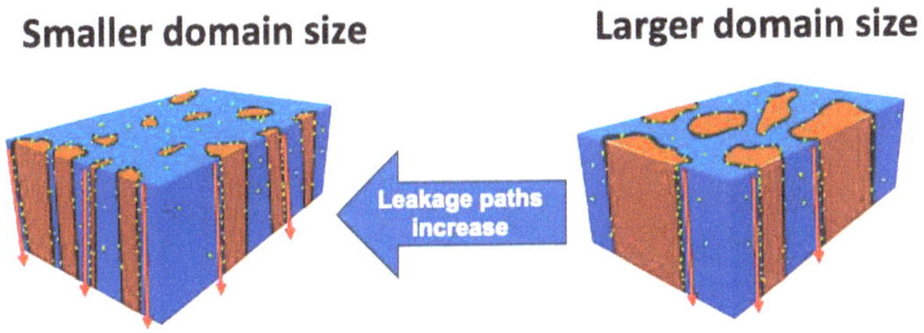

Figure 8. Schematic illustration of domain size's effect on the leakage path.

4. Conclusions

In this study, the impact of domain size on the constant voltage stress TDDB and cycle-to-breakdown reliability are systematically reported. Firstly, MFM devices were fabricated with two different dopants in the ferroelectric layer, which intentionally induced different domain sizes. The PFM analysis indicated that Sample B (Si-doped) had a larger domain size than Sample A (Y-doped). Furthermore, CVS TDDB and cycle-to-breakdown evaluations were conducted in Sample A and Sample B, indicating that Sample B had better CVS TDDB and cycle-to-breakdown stability. A clear correlation was observed between the larger domain size and better time-dependent stability, which may be attributed to fewer domain boundaries in Sample B. We are the first to report the effects of domain size on CVS TDDB and cycle-to-breakdown reliability and conclude that optimizing the domain size can improve devices' reliability.

Author Contributions: Conceptualization, T.-Y.C., K.-C.W. and T.-L.W.; methodology, T.-Y.C., K.-C.W. and T.-L.W.; formal analysis, T.-Y.C., K.-C.W., H.-Y.L., N.R., U.C. and T.-L.W.; investigation, T.-Y.C., K.-C.W., H.-Y.L., N.R., U.C. and T.-L.W.; data curation, T.-Y.C. and K.-C.W.; writing—original draft preparation, T.-Y.C. and T.-L.W.; writing—review and editing, J.-H.H., W.-C.P., N.R., U.C., K.B., J.V.H. and T.-L.W.; supervision, N.R. and T.-L.W.; project administration, N.R. and T.-L.W.; funding acquisition, T.-L.W. All authors have read and agreed to the published version of the manuscript.

Funding: This work was financially supported by the "Advanced Semiconductor Technology Research Center" from The Featured Areas Research Center Program within the framework of the Higher Education Sprout Project by the Ministry of Education (MOE) in Taiwan. This work was also partly supported by the Ministry of Science and Technology, Taiwan, under Grant MOST 111-2634-F-A49-008, 111-2622-8-A49-018-SB, and 112-2636-E-A49-005.

Institutional Review Board Statement: Not applicable.

Informed Consent Statement: Not applicable.

Data Availability Statement: The data presented in this study are available upon request from the corresponding author. The data are not publicly available due to privacy.

Acknowledgments: This work was financially supported by the "Advanced Semiconductor Technology Research Center" from The Featured Areas Research Center Program within the framework of the Higher Education Sprout Project by the Ministry of Education (MOE) in Taiwan. This work was also partly supported by the Ministry of Science and Technology, Taiwan, under Grant MOST 111-2634-F-A49-008, 111-2622-8-A49-018-SB, and 112-2636-E-A49-005.

Conflicts of Interest: The authors declare no conflict of interest.

References

1. Böscke, T.S.; Müller, J.; Bräuhaus, D.; Schröder, U.; Böttger, U. Ferroelectricity in hafnium oxide thin films. *Appl. Phys. Lett.* **2011**, *99*, 102903. [CrossRef]
2. Song, C.M.; Kwon, H.J. Ferroelectrics based on HfO_2 film. *Electronics* **2021**, *10*, 2759. [CrossRef]
3. Müller, J.; Böscke, T.S.; Schröder, U.; Mueller, S.; Bräuhaus, D.; Böttger, U.; Frey, L.; Mikolajick, T. Ferroelectricity in Simple Binary ZrO_2 and HfO_2. *Nano Lett.* **2012**, *12*, 4318–4323. [CrossRef] [PubMed]
4. Cheema, S.S.; Kwon, D.; Shanker, N.; Reis, R.D.; Hsu, S.-L.; Xiao, J.; Zhang, H.; Wagner, R.; Datar, A.; McCarter, M.R.; et al. Enhanced ferroelectricity in ultrathin films grown directly on silicon. *Nature* **2020**, *580*, 478–482. [CrossRef] [PubMed]
5. Banerjee, W. Challenges and Applications of Emerging Nonvolatile Memory Devices. *Electronics* **2020**, *9*, 1029. [CrossRef]
6. Salahuddin, S.; Datta, S. Use of negative capacitance to provide voltage amplification for low power nanoscale devices. *Nano Lett.* **2008**, *8*, 405–410. [CrossRef] [PubMed]
7. Chou, C.-P.; Lin, Y.-X.; Huang, Y.-K.; Chan, C.-Y.; Wu, Y.-H. Junctionless Poly-GeSn Ferroelectric Thin-Film Transistors with Improved Reliability by Interface Engineering for Neuromorphic Computing. *ACS Appl. Mater. Interfaces* **2019**, *12*, 1014–1023. [CrossRef] [PubMed]
8. MKim, M.K.; Lee, J.S. Ferroelectric analog synaptic transistors. *Nano Lett.* **2019**, *19*, 2044–2050. [CrossRef]
9. Jerry, M.; Chen, P.Y.; Zhang, J.; Sharma, P.; Ni, K.; Yu, S.; Datta, S. Ferroelectric FET analog synapse for acceleration of deep neural network training. In Proceedings of the 2017 IEEE International Electron Devices Meeting (IEDM), San Francisco, CA, USA, 2–6 December 2017; pp. 6.2.1–6.2.4. [CrossRef]
10. Kimura, H.; Hanyu, T.; Kameyama, M.; Fujimori, Y.; Nakamura, T.; Takasu, H. Complementary ferroelectric-capacitor logic for low-power logic-in-memory VLSI. *IEEE J. Solid-State Circuits* **2004**, *39*, 919–926. [CrossRef]
11. Wang, Y.Y.; Wang, K.C.; Wu, C.H.; Chang, T.Y.; Ronchi, N.; Banerjee, K.; Van den Bosch, G.; Van Houdt, J.; Wu, T.L. Demonstration of 64 Conductance States and Large Dynamic Range in Si-doped HfO_2 FeFETs under Neuromorphic Computing Operations. In Proceedings of the 2022 International Symposium on VLSI Technology, Systems and Applications (VLSI-TSA), Hsinchu, Taiwan, 18–21 April 2022; pp. 1–2. [CrossRef]
12. Chen, Y.H.; Su, C.J.; Hu, C.; Wu, T.L. Effects of Annealing on Ferroelectric Hafnium Zirconium Oxide-Based Transistor Technology. *IEEE Electron Device Lett.* **2019**, *40*, 467–470. [CrossRef]
13. Cao, R.; Wang, Y.; Zhao, S.; Yang, Y.; Zhao, X.; Wang, W.; Zhang, X.; Lv, H.; Liu, Q.; Liu, M. Effects of Capping Electrode on Ferroelectric Properties of $Hf_{0.5}Zr_{0.5}O_2$ Thin Films. *IEEE Electron Device Lett.* **2018**, *39*, 1207–1210. [CrossRef]
14. Ronchi, N.; Ragnarsson, L.Å.; Breuil, L.; Banerjee, K.; McMitchell, S.R.C.; O'Sullivan, B.; Milenin, A.; Kundu, S.; Pak, M.; Van Houdt, J. Ferroelectric FET with Gd-doped HfO_2: A Step Towards Better Uniformity and Improved Memory Performance. In Proceedings of the 2021 Silicon Nanoelectronics Workshop (SNW), Kyoto, Japan, 13 June 2021; pp. 1–2.
15. Skopina, E.V.; Guillaume, N.; Alrifai, L.; Gonon, P.; Bsiesya, A. Sub-10-nm ferroelectric Gd-doped HfO_2 layers. *Appl. Phys. Lett.* **2022**, *120*, 172901. [CrossRef]
16. Tashiro, Y.; Shimizu, T.; Mimura, T.; Funakubo, H. Comprehensive Study on the Kinetic Formation of the Orthorhombic Ferroelectric Phase in Epitaxial Y-Doped Ferroelectric HfO_2 Thin Films. *ACS Appl. Electron. Mater.* **2021**, *3*, 3123–3130. [CrossRef]
17. Li, S.; Zhou, D.; Shi, Z.; Hoffmann, M.; Mikolajick, T.; Schroeder, U. Temperature-Dependent Subcycling Behavior of Si-Doped HfO_2 Ferroelectric Thin Films. *ACS Appl. Electron. Mater.* **2021**, *3*, 2415–2422. [CrossRef]
18. Wu, C.-H.; Wang, K.-C.; Wang, Y.-Y.; Hu, C.; Su, C.-J.; Wu, T.-L. Enhancement of Ferroelectricity in 5 nm Metal-Ferroelectric -Insulator Technologies by Using a Strained TiN Electrode. *Nanomaterials* **2022**, *12*, 468. [CrossRef]
19. Wu, C.-H.; Ronchi, N.; Wang, K.-C.; Wang, Y.-Y.; McMitchel, S.; Banerjee, K.; Van den Bosch, G.; Van Houdt, J.; Wu, T.-L. Impacts of Pulsing Schemes on the Endurance of Ferroelectric Metal–Ferroelectric–Insulator–Semiconductor Capacitors. *IEEE J. Electron Devices Soc.* **2022**, *10*, 109–114. [CrossRef]
20. Popovici, M.I.; Bizindavyi, J.; Favia, P.; Clima, S.; Alam, M.N.K.; Ramachandran, R.K.; Walke, A.M.; Celano, U.; Leonhardt, A.; Mukherjee, S.; et al. High performance La-doped HZO based ferroelectric capacitors by interfacial engineering. In Proceedings of the 2022 International Electron Devices Meeting (IEDM), San Francisco, CA, USA, 3–7 December 2022.
21. Tan, A.J.; Liao, Y.-H.; Wang, L. C.; Shanker, N.; Bae, J.-H.; Hu, C.; Salahuddin, S. Ferroelectric HfO_2 Memory Transistors with High-κ Interfacial Layer and Write Endurance Exceeding 1010 Cycles. *IEEE Electron Device Lett.* **2021**, *42*, 994–997. [CrossRef]
22. Zagni, N.; Puglisi, F.M.; Pavan, P.; Alam, M.A. Reliability of HfO_2-Based Ferroelectric FETs: A Critical Review of Current and Future Challenges. *Proc. IEEE* **2023**, *111*, 158–184. [CrossRef]
23. Ronchi, N.; Ragnarsson, L.Å.; Celano, U.; Kaczer, B.; Kaczmarek, K.; Banerjee, K.; McMitchell, S.R.C.; Van Houdt, J. A comprehensive variability study of doped HfO_2 FeFET for memory applications. In Proceedings of the 2022 IEEE International Memory Workshop (IMW), Dresden, Germany, 15–18 May 2022.
24. Wei, X.K.; Sluka, T.; Fraygola, B.; Feigl, L.; Du, H.; Jin, L.; Setter, N. Controlled charging of ferroelastic domain walls in oxide ferroelectrics. *ACS Appl. Mater. Interfaces* **2017**, *9*, 6539–6546. [CrossRef] [PubMed]

nanomaterials

Article

Pt Modified Sb_2Te_3 Alloy Ensuring High−Performance Phase Change Memory

Yang Qiao [1], Jin Zhao [2,3], Haodong Sun [1], Zhitang Song [2], Yuan Xue [2,*], Jiao Li [1,4,*] and Sannian Song [2,*]

[1] The Microelectronic Research & Development Center, Shanghai University, Shanghai 200444, China; yangqiao@shu.edu.cn (Y.Q.); shd_1013484830@shu.edu.cn (H.S.)
[2] State Key Laboratory of Functional Materials for Informatics, Shanghai Institute of Microsystem and Information, Chinese Academy of Sciences, Shanghai 200050, China; zhaojin@mail.sim.ac.cn (J.Z.); ztsong@mail.sim.ac.cn (Z.S.)
[3] University of Chinese Academy of Sciences, Beijing 100049, China
[4] Department of Mechatronics Engineering and Automation, Shanghai University, Shanghai 200444, China
* Correspondence: xueyuan@mail.sim.ac.cn (Y.X.); lijiaoshu@shu.edu.cn (J.L.); songsannian@mail.sim.ac.cn (S.S.)

Abstract: Phase change memory (PCM), due to the advantages in capacity and endurance, has the opportunity to become the next generation of general−purpose memory. However, operation speed and data retention are still bottlenecks for PCM development. The most direct way to solve this problem is to find a material with high speed and good thermal stability. In this paper, platinum doping is proposed to improve performance. The 10-year data retention temperature of the doped material is up to 104 °C; the device achieves an operation speed of 6 ns and more than 3×10^5 operation cycles. An excellent performance was derived from the reduced grain size (10 nm) and the smaller density change rate (4.76%), which are less than those of $Ge_2Sb_2Te_5$ (GST) and Sb_2Te_3. Hence, platinum doping is an effective approach to improve the performance of PCM and provide both good thermal stability and high operation speed.

Keywords: phase change memory; phase change material; high speed; thermal stability

1. Introduction

In the past decades, rapid advances in artificial intelligence [1,2], supercomputing [3], and big data [4] have required ever−faster data exchange. While traditional hard disk drives and solid−state drives struggle to meet demand, new types of memory have taken the challenge. Phase change memory (PCM) is considered a promising non−volatile memory technology due to its advantages of high speed, high density, high scalability, low operating voltage, and high endurance [2,5–8]. As the storage medium of PCM, phase change material can achieve reversible phase transitions between crystalline and amorphous states under the action of electrical pulses. The memory relies on the resistance difference between the crystalline and amorphous states of phase change materials to store "0" and "1" [9–12]. The common phase change material $Ge_2Sb_2Te_5$ (GST) is currently the most successful commercialized material. However, poor 10−year data retention (~85 °C), slow operating speed (~20 ns), and a density change rate of 6.5% limit its wider application in electrical devices [13,14]. Therefore, looking for a phase change material with high amorphous thermal stability and fast speed is the key to improving the performance of PCM [2,15,16].

The PCM device based on Sb_2Te_3 shows fast operation speed. However, the low crystallization temperature (<100 °C) makes the amorphous state unstable, which means that Sb_2Te_3 is not suitable for PCM application. Doping is a good way to improve thermal stability and speed. Some researchers have obtained high−performance phase change materials by doping Sb_2Te_3, such as $Sc_{0.2}Sb_2Te_3$. It achieved an ultra−fast operation speed of 700 ps and the data retention of ~87 °C [5], which satisfies the requirements of subnanosecond high−speed cache memory. However, in some applications filed [13,14], higher data retention is required. Since the thermal stability of materials is related to data retention, we need to find a phase change material with high thermal stability. The traditional precious metals materials (Au, Ag, Pt) have excellent chemical stability and are conducive to engineering applications. Our selection principle is that the element with high electronegativity is used as the doped element, so as to form a stable chemical bond with the elements of the parent material to ensure no phase separation during the operation of the device. Silver was, therefore, rejected as a candidate material. At the same time, considering the cost of gold and platinum, platinum is finally selected as the dopant.

In this work, we have performed electrical tests based on Pt−Sb_2Te_3 devices and microscopic characterization of films. The PCM devices based on $Pt_{0.14}Sb_2Te_3$ (PST) show fast operation speed, high data retention, and good endurance. Meanwhile, the corresponding microstructure of PST explains the origin of its high performance.

2. Materials and Methods

2.1. Film Preparation and Testing

The Sb_2Te_3, $Pt_{0.1}Sb_2Te_3$, $Pt_{0.14}Sb_2Te_3$ (PST), and $Pt_{0.22}Sb_2Te_3$ films are deposited by sputtering of Pt and Sb_2Te_3 targets. The compositions of these films were measured by energy−dispersive spectroscopy (EDS). Films with a thickness of 200 nm were deposited on SiO_2/Si (100) substrates for resistance−temperature (R−T) and X−ray diffraction (XRD) tests. In situ R−T measurement was conducted by a homemade vacuum heating table, and the heating rate was 20 °C /min. The film was heated in a vacuum chamber with a heating rate of 60 °C/min, and the isothermal change in resistance with increasing temperature was recorded to estimate the 10−year data retention. The X−ray reflectivity (XRR) experiment (Bruker D8 Discover) was used to test the density change of films before and after crystallization. X−ray photoelectron spectroscopy (XPS) experiment was used to evaluate the bonding situation. Then the film (about 20 nm) was deposited on the ultra−thin carbon film, and its microstructure was studied by Transmission Electron Microscope (TEM). TEM is manufactured by Hitachi Limited in Tokyo, Japan.

2.2. Device Fabrication

T−shaped PCM devices were prepared by 0.13 μm complementary metal−oxide semiconductor technology. The diameter of the tungsten bottom electrode is about 60 nm. The 70 nm−thick phase change material and 20 nm−thick TiN as adhesion layer were deposited through the sputtering method over a 60 nm diameter of tungsten heating electrode. The device is measured by the Keithley 2400 C source meter and Tektronix AWG5002B pulse generator. The Keithley 2400 C source meter and Tektronix AWG5002B pulse generator are manufactured in the Beaverton, OR, United States by Tektronix.

3. Results

3.1. Improved Device Performance

The films with different Pt compositions were performed by resistance−temperature (R−T) tests, as shown in Figure 1b. The R−T curves show that doping Pt into Sb_2Te_3 can enhance the crystallization temperature of the material, and the crystallization temperature increases with more Pt. The amorphous resistance of the material first increases and then decreases with the content of Pt. This is due to the low crystallization temperature of as−deposited Sb_2Te_3 film and partly crystallization, which will be confirmed by subsequent XRD experiments. Dopant atoms can increase scattering probability, so the effect of scattering is enhanced as the doping concentration increases and results in an increase in resistivity. However, when the doping concentration is too high, the metallicity of the material increases and the resistivity decreases. The crystallization temperature can be measured via Raman or XRD measurements and is simply approximated by the curve of resistivity. In this paper, we chose to use the R−T curve to calculate the crystallization temperature. In the R−T diagram, the crystallization temperatures of $Pt_{0.1}Sb_2Te_3$, $Pt_{0.14}Sb_2Te_3$ (PST), and $Pt_{0.22}Sb_2Te_3$ are 137 °C, 199 °C, and 236 °C, respectively, which indicates that the thermal stability of the Sb_2Te_3 alloy is improved after Pt doping. The resistance of the PST drops by more than an order of magnitude, which is enough to distinguish the ON/OFF states used in the PCM storage devices. Therefore, we believe that the performance of the PST film is greatly improved. Figure 1c shows the resistance time (R−T) curve. The 10−year data retention can be estimated by the Arrhenius equation:

$$t = \tau exp(E_a/K_B T) \quad (1)$$

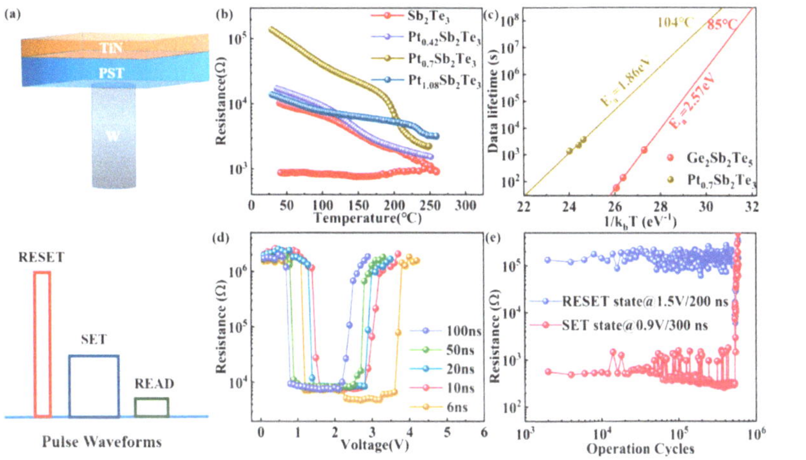

Figure 1. Device performance. (**a**) The schematic diagram of the T−shaped phase change memory (PCM) device. Schematic diagram of three pulse voltages of RESET, SET, and READ of PCM. (**b**) The temperature dependence of the resistance of Sb_2Te_3, $Pt_{0.1}Sb_2Te_3$, $Pt_{0.14}Sb_2Te_3$ (PST), and $Pt_{0.22}Sb_2Te_3$ films at the same heating rate of 20 °C/min. (**c**) At the heating rate of 60 °C/min, the extrapolated fitting line based on the Arrhenius formula shows the 10−year data retention temperature and crystallization activation energy. (**d**) Resistance−voltage characteristics of PST based T−shaped PCM device. The SET−RESET programming windows are obtained under different pulse widths. (**e**) Endurance characteristic of PST based PCM T−shaped devices.

The 10−year data retention for GST and PST are expected to be 85 °C and 104 °C, respectively, with corresponding activation energies (E_a) of 2.57 eV and 1.86 eV. The activation energy errors are 0.05 eV and 0.40 eV, respectively. We find that 10−year data retention of PST films is higher than that of most phase change memories, such as GST (~85 °C) and SST (~87 °C) [5].

Accordingly, based on standard 0.13 μm complementary metal−oxide semiconductor (CMOS) technology, T−shape PCM devices based on PST were fabricated, as shown in Figure 1a. Then, the electrical properties of the device are characterized. RESET, SET, and READ functions can be realized by using different pulse waveforms. Figure 1d shows the SET−RESET windows using the resistance−voltage (R−V) curves. The high/low resistance ratio (R_{RESET}/R_{SET}) is about two orders of magnitude, which can meet the requirement of the ON/OFF ratio used in PCM. When the voltage pulse width of 6 ns, the SET/RESET voltage of the PST device requires 1.2 V/3.8 V. However, GST requires 4.6 V/5.5 V with a 10 ns operation speed [5]. A pre−program voltage applied by pre−operation to GST enables a SET speed of 500 ps in a restricted device structure [17]. This competitive recording speed is already comparable to DRAM and SRAM (1−10 ns) [18]. As shown in Figure 1e, the endurance period is revealed after we alternately apply two appropriate SET and RESET voltage pulses. Figure 1e shows that the reversible phase transition characteristic is up to 5×10^5 switching cycles with a resistance ratio of two orders of magnitude. The switching cycles and resistance ratio of PST are better than Sb_2Te_3 [19]. The endurance performance is higher than GST [20] using the T−shaped device structure. All above, compared with GST, faster operation speed and better endurance of PST have proved Pt doping Sb_2Te_3 with suitable composition is a promising novel phase−change material.

3.2. Characterization of Thin Film Structure

The XRD method was employed to characterize the lattice structure of PST film. Figure 2a,b shows the XRD results of PST and Sb_2Te_3 films at different annealing temperatures. The diffraction peak of Sb_2Te_3 appears in the deposited state, indicating that the deposited Sb_2Te_3 has crystallized. At this time, there is no diffraction peak of PST, so the PST has not crystallized. At 200 °C, the FCC phase appeared in the PST, which indicated that Pt inhibited the formation of the FCC phase and increased the crystallization temperature. When the annealing temperature is 260 °C, both PST and Sb_2Te_3 have only the diffraction peaks of the hexagonal phase. Compared with pure Sb_2Te_3 film, the diffraction peaks of PST film become wider, the intensity of the peak becomes lower, and some diffraction peaks disappear. In addition, a difference in the full width at half maximum (FWHM) of the diffraction peak is observed on the XRD curves. According to the Scherrer formula:

$$\beta = K\lambda / L(cos\theta) \tag{2}$$

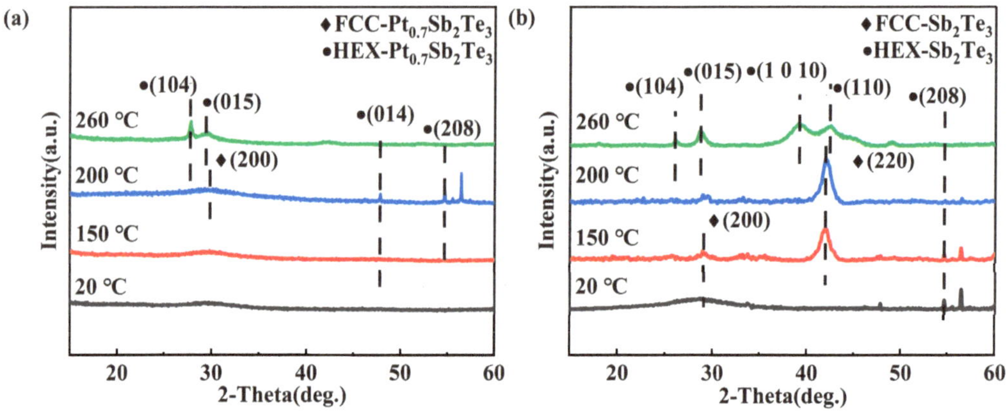

Figure 2. XRD results of the Sb_2Te_3 and PST. (**a**,**b**) XRD curves of PST and Sb_2Te_3 films were annealed at 150 °C, 200 °C, and 260 °C for 5 min in an N_2 atmosphere.

K in the equation is the Scherrer constant (K = 0.89), β is the grain size, L is the full width at half maximum (FWHM) of the diffraction peak of the sample, θ is the diffraction angle, and λ is the X−ray wavelength (0.154056 nm). The FWHM of PST was significantly higher than that of Sb_2Te_3, indicating that the incorporation of Pt inhibited the crystallization growth process, and grain refinement was obvious. Reducing grain size is ideal for programming areas [21].

To study the crystalline phase and grain size more intuitively, high−resolution transmission electron microscopy (HRTEM) and the associated selected area electron diffraction (SAED) patterns for Sb_2Te_3 film and PST films are presented in Figure 3. In total, 2 samples were annealed at the temperature of 260 °C for 5 min. The annealed films are both in a polycrystalline state. Comparing Figure 3a,b, it can clearly be seen that the grain size decreases significantly. In Figure 3c, doping Pt reduces the grain size of Sb_2Te_3 from 50 nm to about 5~10 nm, which confirms that the half−height width of PST is much larger than that of Sb_2Te_3. Meanwhile, according to Figure 3a,b, small crystal grains of the PST film can be also inferred from the continuous diffraction rings [22]. Smaller grain size increases the surface volume ratio, thus generating more grain boundaries [23]. As the number of grain boundaries increases, the crystal diffusion and slippage can be reduced. Hence, the residual stress in the bulk of films can be degraded [24,25]. Moreover, the increased grain boundaries provide a phonon and electron scattering center, and the decreased thermal and electrical conductivity will improve the energy efficiency of the Joule heating [26]. According to the HRTEM image in Figure 3d,e, the crystal structure is in the hexagonal phase after the calculation of inter−planar distance. They all belong to the (1010) and (105) families, which indicates the crystalline state of the PST film is composed of the hexagonal phase. The result of SAED in Figure 3b matches the HRTEM perfectly. In other words, Pt doping affects the crystallization behavior of the Sb_2Te_3 film without forming any new phase or structure.

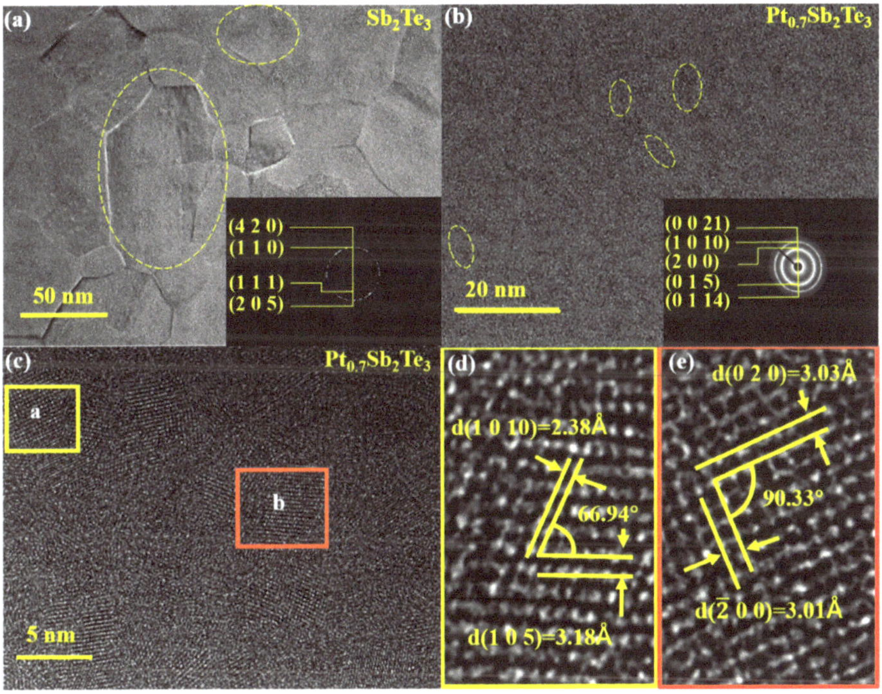

Figure 3. (**a**) TEM image of Sb_2Te_3 film after annealed at 260 °C. (**b**) TEM image of PST film after annealed at 260 °C. (**c**–**e**) HRTEM images of PST film after annealed at 260 °C.

Crystallization usually leads to an increase in film density and a reduction in film thickness. The information on the density change upon crystallization is of paramount importance in phase change media technology since it is related to the stresses induced in the system during the write/erase cycle. The change of density before and after the phase transition of the sample was measured by XRR. Figure 4 separately depicts the XRR curves of PST films in amorphous and crystalline states. Based on the peak position shift, a linear fit calculation is performed, as shown in Figure 4b. During the transition from the amorphous to the crystalline state, the thickness change rate of PST film is only 4.7%, while the thickness change rate of Sb_2Te_3 and GST films are 7.5% [27] and 6.5%, respectively. This enhancement is responsible for the improved cyclability.

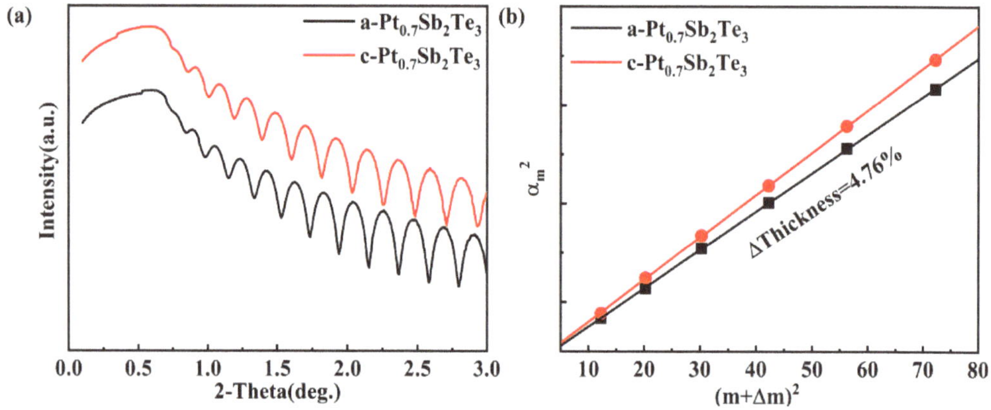

Figure 4. The density−change rate before and after PST crystallization (**a**) XRR curves of amorphous and crystalline PST films. (**b**) Bragg fitting curves of amorphous and crystalline films.

3.3. Evidence of Pt Occupying Positions

Experiments have proved that when element B is replaced by element C and bonded with element A, if the electronegativity of element C is greater than that of element B, the binding energy of element A increases [28]. In Figure 5a,b, the binding state of Sb_2Te_3 and PST is revealed by XPS. When the Pt atom enters Sb_2Te_3, if the Pt atom replaces the Sb atom and combines with the Te atom, since the electronegativity of Pt (2.2) is higher than that of Sb (2.05) and Te (2.12), the binding energy of Te will shift towards the high binding energy, which is consistent with the phenomenon in the experiment in Figure 5. Combined with the XRD result that shows there is no new phase, this confirms that Pt replaces the position of Sb.

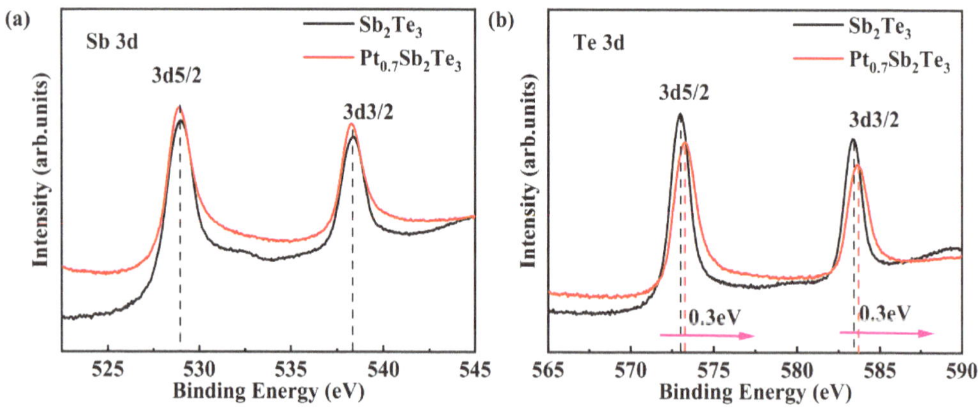

Figure 5. XPS spectra of Sb_2Te_3 and PST films annealed at 260 °C (**a**) Sb 3d and (**b**) Te 3d.

4. Conclusions

In this work, we systematically studied the performance of PST. The PCM devices based on PST can achieve higher speed and data retention than GST devices. According to XPS and TEM analyses, the microstructure feature of Pt−modification Sb_2Te_3 film is explained clearly. The reduced grain size and formation of Pt−Te bonds are the main reasons for the improved properties. Subsequently, a boost in device endurance gave the credit to the reduced density change rate. The improvement of these properties is conducive to the commercial application of the material. Such experimental results show that PST has broad application prospects in complex environments.

Author Contributions: Conceptualization, S.S., Z.S., and Y.X.; methodology, Y.Q., Y.X. and J.Z.; formal analysis, investigation, data curation, and writing—original draft preparation, Y.Q.; writing—review and editing, Y.X., S.S., J.Z., H.S. and J.L.; visualization, Y.X.; supervision, Y.X., S.S. and J.L.; project administration, S.S. and Z.S.; funding acquisition, S.S. and Z.S. All authors have read and agreed to the published version of the manuscript.

Funding: This research was funded by the National Key Research and Development Program of China (2017YFA0206101), Strategic Priority Research Program of the Chinese Academy of Sciences (XDB44010200), National Natural Science Foundation of China (91964204, 61874129, 61874178, 61775008, 61904046), Science and Technology Council of Shanghai (20501120300, 19JC1416800, 19YF1456100), Shanghai Sailing Program (19YF1456100), fund of the State Key Laboratory of Advanced Technologies for Comprehensive Utilization of Platinum Metals, Genetic Engineering of Precious Metal Materials in Yunnan Province (I)–Construction and Application of Precious Metal Materials Professional Database (I) (202002AB080001–1).

Institutional Review Board Statement: Not applicable.

Informed Consent Statement: Not applicable.

Data Availability Statement: Not applicable.

Acknowledgments: The authors acknowledge the financial support from the State Key Laboratory of Functional Materials for Informatics, Shanghai Institute of Microsystem and Information, Chinese Academy of Sciences, and the Microelectronic Research & Development Center, Shanghai University.

Conflicts of Interest: The authors declare no conflict of interest.

References

1. Feldmann, J.; Youngblood, N.; Wright, C.D.; Bhaskaran, H.; Pernice, W.H.P. All-optical spiking neurosynaptic networks with self-learning capabilities. *Nature* **2019**, *569*, 208–214. [CrossRef] [PubMed]
2. Zhang, W.; Mazzarello, R.; Wuttig, M.; Ma, E. Designing crystallization in phase-change materials for universal memory and neuro-inspired computing. *Nat. Rev. Mater.* **2019**, *4*, 150–168. [CrossRef]
3. Ding, K.Y.; Wang, J.J.; Zhou, Y.X.; Tian, H.; Lu, L.; Mazzarello, R.; Jia, C.L.; Zhang, W.; Rao, F.; Ma, E. Phase-change heterostructure enables ultralow noise and drift for memory operation. *Science* **2019**, *366*, 210–215. [CrossRef] [PubMed]
4. Big Data Needs a Hardware Revolution. Available online: https://www.nature.com/articles/d41586-018-01683-1 (accessed on 15 May 2022).
5. Rao, F.; Ding, K.Y.; Zhou, Y.X.; Zheng, Y.H.; Xia, M.J.; Lv, S.L.; Song, Z.T.; Feng, S.L.; Ronneberger, I.; Mazzarello, R.; et al. Reducing the stochasticity of crystal nucleation to enable subnanosecond memory writing. *Science* **2017**, *358*, 1423–1426. [CrossRef] [PubMed]
6. Zhu, M.; Song, W.; Konze, P.M.; Li, T.; Gault, B.; Chen, X.; Shen, J.; Lv, S.; Song, Z.; Wuttig, M.; et al. Direct atomic insight into the role of dopants in phase-change materials. *Nat. Commun.* **2019**, *10*, 3525. [CrossRef] [PubMed]
7. Zhao, J.; Song, W.-X.; Xin, T.; Song, Z. Rules of hierarchical melt and coordinate bond to design crystallization in doped phase change materials. *Nat. Commun.* **2021**, *12*, 6473. [CrossRef]
8. Xue, Y.; Cheng, Y.; Zheng, Y.; Yan, S.; Song, W.; Lv, S.; Song, S.; Song, Z. Phase change memory based on Ta–Sb–Te alloy –Towards a universal memory. *Mater. Today Phys.* **2020**, *15*, 100266. [CrossRef]
9. Zhou, X.; Wu, L.; Song, Z.; Cheng, Y.; Rao, F.; Ren, K.; Song, S.; Liu, B.; Feng, S. Nitrogen-doped Sb-rich Si–Sb–Te phase-change material for high-performance phase-change memory. *Acta Mater.* **2013**, *61*, 7324–7333. [CrossRef]
10. Simpson, R.E. The changing phase of data storage. *Nat. Nanotechnol.* **2019**, *14*, 643–644. [CrossRef]
11. Wuttig, M.; Yamada, N. Phase-change materials for rewriteable data storage. *Nat. Mater.* **2007**, *6*, 824–832. [CrossRef]
12. Xue, Y.; Song, S.; Chen, X.; Yan, S.; Lv, S.; Xin, T.; Song, Z. Enhanced performance of phase change memory by grain size reduction. *J. Mater. Chem. C* **2022**, *10*, 3585–3592. [CrossRef]
13. Njoroge, W.K.; Woltgens, H.W.; Wuttig, M. Density changes upon crystallization of Ge2Sb2.04Te4.74 films. *J. Vac. Sci. Technol. A-Vac. Surf. Film.* **2002**, *20*, 230–233. [CrossRef]
14. Guo, T.Q.; Song, S.N.; Song, Z.T.; Ji, X.L.; Xue, Y.; Chen, L.L.; Cheng, Y.; Liu, B.; Wu, L.C.; Qi, M.; et al. SiC-Doped Ge2Sb2Te5 Phase-Change Material: A Candidate for High-Density Embedded Memory Application. *Adv. Electron. Mater.* **2018**, *4*, 1800083. [CrossRef]
15. Fong, S.W.; Neumann, C.M.; Wong, H.S.P. Phase-Change Memory—Towards a Storage-Class Memory. *IEEE Trans. Electron. Devices* **2017**, *64*, 4374–4385. [CrossRef]

16. Wang, Z.R.; Wu, H.Q.; Burr, G.W.; Hwang, C.S.; Wang, K.L.; Xia, Q.F.; Yang, J.J. Resistive switching materials for information processing. *Nat. Rev. Mater.* **2020**, *5*, 173–195. [CrossRef]
17. Loke, D.; Lee, T.H.; Wang, W.J.; Shi, L.P.; Zhao, R.; Yeo, Y.C.; Chong, T.C.; Elliott, S.R. Breaking the Speed Limits of Phase-Change Memory. *Science* **2012**, *336*, 1566–1569. [CrossRef]
18. Wong, H.S.; Salahuddin, S. Memory leads the way to better computing. *Nat. Nanotechnol.* **2015**, *10*, 191–194. [CrossRef]
19. Hwang, S.; Park, H.; Kim, D.; Lim, H.; Lee, C.; Han, J.H.; Kwon, Y.K.; Cho, M.H. Ultra-low Energy Phase Change Memory with Improved Thermal Stability by Tailoring the Local Structure through Ag Doping. *ACS Appl. Mater. Interfaces* **2020**, *12*, 37285–37294. [CrossRef]
20. Cubukcu, M.; Venkateshvaran, D.; Wittmann, A.; Wang, S.-J.; Di Pietro, R.; Auffret, S.; Vila, L.; Wunderlich, J.; Sirringhaus, H. Electrical nucleation and detection of single 360° homochiral Néel domain walls measured using the anomalous Nernst effect. *Appl. Phys. Lett.* **2018**, *112*, 262409. [CrossRef]
21. Wang, W.J.; Loke, D.; Law, L.T.; Shi, L.P.; Zhao, R.; Li, M.H.; Chen, L.L.; Yang, H.X.; Yeo, Y.C.; Adeyeye, A.O.; et al. Engineering Grains of $Ge_2Sb_2Te_5$ for Realizing Fast-Speed, Low-Power, and Low-Drift Phase-Change Memories with Further Multilevel Capabilities. In Proceedings of the IEEE International Electron Devices Meeting (IEDM), San Francisco, CA, USA, 10–13 December 2012.
22. Cheng, Y.; Song, Z.T.; Gu, Y.F.; Song, S.N.; Rao, F.; Wu, L.C.; Liu, B.; Feng, S.L. Influence of silicon on the thermally-induced crystallization process of Si-Sb_4Te phase change materials. *Appl. Phys. Lett.* **2011**, *99*, 261914. [CrossRef]
23. Penn, R.L.; Banfield, J.F. Imperfect oriented attachment: Dislocation generation in defect-free nanocrystals. *Science* **1998**, *281*, 969–971. [CrossRef] [PubMed]
24. Rao, F.; Song, Z.T.; Ren, K.; Zhou, X.L.; Cheng, Y.; Wu, L.C.; Liu, B. Si-Sb-Te materials for phase change memory applications. *Nanotechnology* **2011**, *22*, 145702. [CrossRef] [PubMed]
25. Zhou, X.L.; Kalikka, J.; Ji, X.L.; Wu, L.C.; Song, Z.T.; Simpson, R.E. Phase-Change Memory Materials by Design: A Strain Engineering Approach. *Adv. Mater.* **2016**, *28*, 3007–3016. [CrossRef] [PubMed]
26. Wang, Y.; Guo, T.Q.; Liu, G.Y.; Li, T.; Lv, S.L.; Song, S.N.; Cheng, Y.; Song, W.X.; Ren, K.; Song, Z.T. Sc-Centered Octahedron Enables High-Speed Phase Change Memory with Improved Data Retention and Reduced Power Consumption. *ACS Appl. Mater. Interfaces* **2019**, *11*, 10848–10855. [CrossRef] [PubMed]
27. Xia, M.; Zhu, M.; Wang, Y.; Song, Z.; Rao, F.; Wu, L.; Cheng, Y.; Song, S. Ti-Sb-Te alloy: A candidate for fast and long-life phase-change memory. *ACS Appl. Mater. Interfaces* **2015**, *7*, 7627–7634. [CrossRef]
28. Zhu, M.; Wu, L.C.; Song, Z.T.; Rao, F.; Cai, D.L.; Peng, C.; Zhou, X.L.; Ren, K.; Song, S.N.; Liu, B.; et al. $Ti_{10}Sb_{60}Te_{30}$ for phase change memory with high-temperature data retention and rapid crystallization speed. *Appl. Phys. Lett.* **2012**, *100*, 122101. [CrossRef]

nanomaterials

MDPI

Article

Study of Selective Dry Etching Effects of 15-Cycle $Si_{0.7}Ge_{0.3}$/Si Multilayer Structure in Gate-All-Around Transistor Process

Enxu Liu [1,2], Junjie Li [1,2,*], Na Zhou [1,2], Rui Chen [1,2,*], Hua Shao [1,2], Jianfeng Gao [1,2], Qingzhu Zhang [1,2], Zhenzhen Kong [1,2], Hongxiao Lin [1,2], Chenchen Zhang [1], Panpan Lai [1,2], Chaoran Yang [1,2], Yang Liu [1], Guilei Wang [3,4,*], Chao Zhao [3], Tao Yang [1,2], Huaxiang Yin [1,2], Junfeng Li [1,2], Jun Luo [1,2] and Wenwu Wang [1,2]

1 Key Laboratory of Microelectronics Devices & Integrated Technology, Institute of Microelectronics of Chinese Academy of Sciences (IMECAS), Beijing 100029, China; liuenxu@ime.ac.cn (E.L.); tyang@ime.ac.cn (T.Y.)
2 Microelectronics Institute, University of Chinese Academy of Sciences, Beijing 100049, China
3 Beijing Superstring Academy of Memory Technology, Beijing 100176, China; chao.zhao@bjsamt.org.cn
4 Hefei National Laboratory, Hefei 230088, China
* Correspondence: lijunjie@ime.ac.cn (J.L.); chenrui1@ime.ac.cn (R.C.); guilei.wang@bjsamt.org.cn (G.W.)

Abstract: Gate-all-around (GAA) structures are important for future logic devices and 3D-DRAM. Inner-spacer cavity etching and channel release both require selective etching of $Si_{0.7}Ge_{0.3}$. Increasing the number of channel-stacking layers is an effective way to improve device current-driving capability and storage density. Previous work investigated ICP selective etching of a three-cycle $Si_{0.7}Ge_{0.3}$/Si multilayer structure and the related etching effects. This study focuses on the dry etching of a 15-cycle $Si_{0.7}Ge_{0.3}$/Si multilayer structure and the associated etching effects, using simulation and experimentation. The simulation predicts the random effect of lateral etching depth and the asymmetric effect of silicon nanosheet damage on the edge, both of which are verified by experiments. Furthermore, the study experimentally investigates the influence and mechanism of pressure, power, and other parameters on the etching results. Research on these etching effects and mechanisms will provide important points of reference for the dry selective etching of $Si_{0.7}Ge_{0.3}$ in GAA structures.

Keywords: GAA; selective etch; silicon germanium; etch effect; process simulation

1. Introduction

In integrated circuit manufacturing, engineers have been working hard to push devices to downscale critical dimensions in accordance with Moore's Law [1]. The advanced technology node has adopted three-dimensional (3D) multilayer in the vertical direction after reaching the limit of integration in the horizontal direction. In terms of logic devices, the MOSFET structure has evolved from the conventional planar structure to the fin structure (FinFET), and then to the gate-all-around (GAA) structure in the past decade [2–4]. In terms of memory devices, 3D stacking can boost memory density and performance while reducing costs. The 3D-NAND relies more on stacked layers to break the bottleneck of memory capacity, the current maximum of which has exceeded 200 layers [5]. For DRAM memory, 3D-DRAM can follow the example of 3D-NAND flash memory by flipping the battery and stacking a large number of layers to meet the shrinking capacitor size. Samsung first demonstrated this structure for 3D-DRAM at the VLSI conference in June 2023 [6]. Since 3D-DRAM requires a large number of layer stacks to satisfy the space consumption in the horizontal direction, research on the etching of ultra-multilayer structures becomes necessary.

In order to achieve a multilayer structure in GAAFET and 3D-DRAM, it is necessary to employ the epitaxy technology of $Si_{1-x}Ge_x$/Si multilayer [7] and the selective etching of $Si_{1-x}Ge_x$ [8]. This paper will focus on the selective etching of $Si_{1-x}Ge_x$. The reason for the etching selectivity of SiGe is that the bond energy of Si–Ge is lower than the bond energy between Si–Si, making it easier to break [8]. Another explanation is that the doping of Si with Ge reduces the activation energy of the reaction, resulting in a reaction rate for $Si_{1-x}Ge_x$ greater than that for Si [9]. At present, conventional methods for $Si_{1-x}Ge_x$ selective etching mainly include wet etching [10], gaseous HCl etching [11], and dry etching [12]. In wet etching, a mixed solution containing H_2O_2, HNO_3, CH_3COOH, and HF is usually adopted to selectively etch the $Si_{1-x}Ge_x$ layer [13,14]. However, there are significant limitations in high-density circuit arrays and nanosheet devices with large aspect ratios. Vapor phase etching using HCl in chemical vapor deposition (CVD) reactors is also limited by its high-temperature environment and crystalline orientation-dependent etching [15]. Dry plasma etching has become a common method for etching $Si_{1-x}Ge_x$ in recent years, usually using plasma containing halogen elements for selective etching. Dry plasma etching mainly uses CF_4 and NF_3 as etching gases, together with Ar, O_2, and He as auxiliary gases. The method can control critical dimensions well and has better etching uniformity, which is more desired in GAA devices and 3D-DRAM [16–20].

In order to obtain higher performance, multiple silicon channel layers can be stacked in the vertical direction [21]. Barraud successfully prepared a stack of seven silicon nanosheets, the driving current of which was three times that of two-cycle-stacked devices [22]. Seven or eight Si nanosheets is the limit reported in the literature so far. Stacks with more nanosheets, such as 10 or even 15 nanosheets, have not been studied in detail. Since SiGe selective-etching technology is an important key process in GAA devices, the articles published in recent years usually only show the process results, without mentioning the details of the process [20]. As for analysis of the etching effects, an article in 2019 analyzed the corner-rounding problem due to the diffusion of germanium elements caused by anisotropic etching, but rarely analyzed the profile after isotropic etching. Therefore, this paper focuses on the effects of multilayer etching and explores the future use of selective etching of SiGe in 3D-DRAM.

In our previous research work, we mainly focused on the $Si_{1-x}Ge_x$ selective etching of a three-cycle multilayer structure using $CF_4/O_2/He$ gas [12]. In the balance process between stacked epitaxial growth and SiGe selective etching, $Si_{0.7}Ge_{0.3}$ is considered the optimal atomic ratio [23]. This study is the first to obtain a good isotropic $Si_{0.7}Ge_{0.3}$ selective-etching process using a conventional inductively coupled plasma (ICP) etching on a 15-cycle multilayer structure, and it also combines simulation with the process to optimize the phenomena that occur during etching and analysis. In a stack of more than a dozen nanosheets, the etching differences and mechanisms at different locations above and below are worth studying and discussing. Simulations by a commercial plasma process simulator named PEGASUS 2022 (PEGASUS Software Inc., Tokyo, Japan) show that random effects in etching are susceptible to the internal environment of the cavity in a multilayer. The edge of Si nanosheets will display asymmetrical damage phenomena on the up and down sides. Through the simulation, we investigated the influence of parameters such as pressure and power on the etching results of multilayers. And, as the number of stacked layers increases, the consistency of etch depth becomes an increasingly important challenge. By adjusting the process, we achieved a consistent etching of the 15-layer structure. Finally, we characterized the microscopic morphology of $Si_{0.7}Ge_{0.3}$ after etching, and discuss the layer quality and strain relaxation of SiGe during the removal process.

2. Materials and Methods

The experiments in this work were performed on 8-inch (100) silicon wafers. The $Si_{0.7}Ge_{0.3}/Si$ multilayer was grown in an ASM E2000plus RPCVD reactor (ASM, Munich, Germany) with 15 cycles of $Si_{0.7}Ge_{0.3}$ (20 nm)/Si (20 nm) at around 650 °C. Before the epitaxial growth, a 200 mm Si (001) substrate was cleaned with DHF (1:100). Then, the wafer was loaded into the reduced-pressure chemical vapor deposition (RPCVD) chamber and baked at 1050 °C for 2 min to remove the native oxide. The main precursors were germane (GeH_4) and dichlorosilane (SiH_2Cl_2), respectively. And the $Si_{0.7}Ge_{0.3}/Si$ multilayer was grown with H_2 as the carrier gas. Then, a hardmask was deposited on top of the epitaxial layers. Photolithography was carried out to define the pattern in the photoresist on top of the hardmask. Both anisotropic and isotropic etching in this experiment were carried out using an 8-inch ICP etching tool Lam9400DFM (Lam Research Inc., Fremont, CA, USA), and hardmask openings were formed using $CF_4/HBr/O_2$ gas. HBr/O_2 plasma was used to vertically etch the stacked $Si_{0.7}Ge_{0.3}/Si$ multilayer film to obtain a mesa structure with a width of 1 μm. Afterwards, the prepared samples were cut into slices of about 3×3 cm^2 to facilitate etching experiments. Finally, we used CF_4, O_2, and He for the selective etching of the $Si_{0.7}Ge_{0.3}$. The temperature was set to 80 °C, the pressure range was 5–80 mT (full scale 80 mT), the source power range was 200–800 W (full scale 800 W), and bias RF was set to 0 W to minimize ion bombardment. The experimental process is shown in Figure 1.

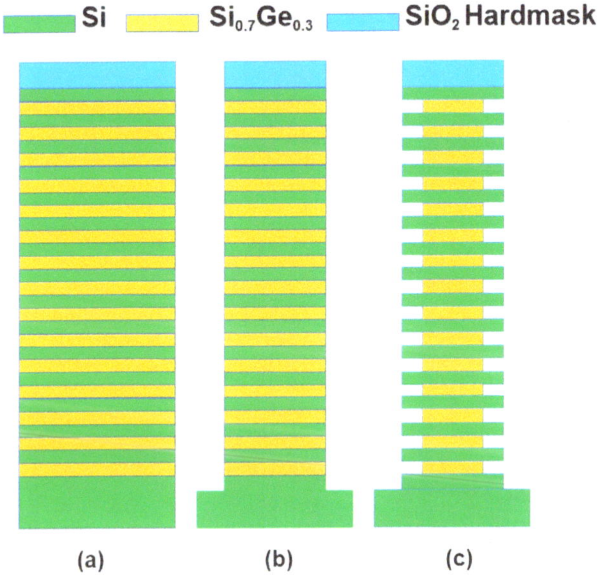

Figure 1. Process flow and schematic diagram of the experiment: (**a**) epitaxy $Si_{0.7}Ge_{0.3}/Si$ multilayer structure; (**b**) photolithographic patterning and dry anisotropic etching; (**c**) dry selective isotropic etching.

PEGASUS 2022 is capable of simulating the time-dependent feature profiles that result from multiple physical and chemical reactions. It employs the Monte Carlo method for the simulation to model the reflection, deposition, etching, and sputtering reactions and the corresponding reaction probabilities. We calculated two-dimensional (2D) profiles of the plasma etching using the feature profile simulation module (FPSM) of the PEGASUS 2022 software. Firstly, the geometry of the substrate to be etched is characterized by the volume occupancy of the solid layer. Second, it is important for the accuracy of our simulations to define as accurately as possible the physical and chemical reactions between the relevant particles (ions and neutral radicals) and the solid layer. When the occupancy becomes 0.0, the cell becomes empty (gas phase), which means it is etched away. This process is repeated until a preset time or all particles are consumed. Finally, the etching profile, gas density, and other important parameters are obtained.

3. Results and Discussion

3.1. Microscopic Characterization of Epitaxial Thin Films

The epitaxy of multicycle $Si_{0.7}Ge_{0.3}/Si$ determines the number of future conductive channels and is the basis for the preparation of channel quality together with channel release etching. In this paper, using epitaxy, 15-cycle multilayer structure of $Si_{0.7}Ge_{0.3}/Si$ was alternately deposited, each layer with a thickness of approximately 20 nm. As shown in Figure 2a, the interface between the $Si_{0.7}Ge_{0.3}$ and the Si is clear, the thickness of each layer is uniform, and the interface profile is straight. Figure 2b is an energy-dispersive spectroscopy (EDS) line-scan diagram of the epitaxial region. The results show that among the 15 $Si_{0.7}Ge_{0.3}$ layers, the Ge content of each layer is constant at around 30%, and the boundaries of each layer are consistent with the growth of the material. They also show that there is no obvious interdiffusion at the interface of different interlayer materials at the junction of the epitaxial stacks. The above results all show that a high-quality 15-cycle $Si_{0.7}Ge_{0.3}$/Si has been prepared by epitaxy, laying the foundation for subsequent selective etching.

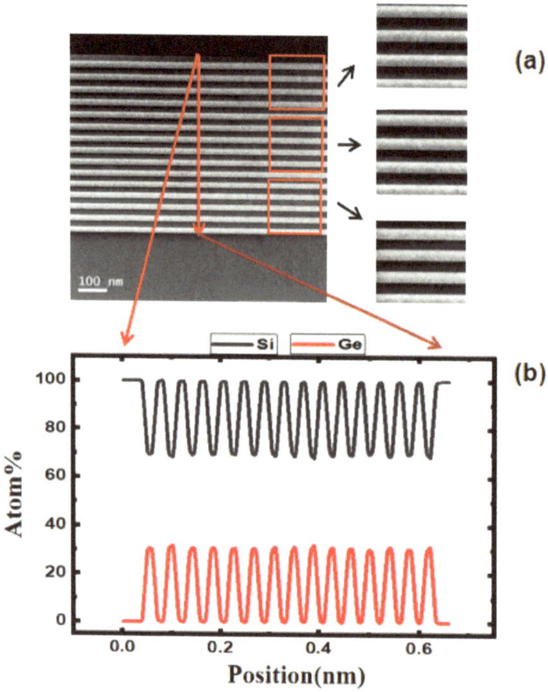

Figure 2. Transmission electron microscope (TEM) image of 15-cycle multilayer structure and EDS analysis: (**a**) high-resolution transmission electron microscope (HRTEM) of 20 nm Si and $Si_{0.7}Ge_{0.3}$ layer; (**b**) EDS line-scanning of filmstack.

3.2. Simulation

PEGASUS 2022 simulation was used to obtain the results of the etching profile under different conditions. By constructing the ICP cavity environment, CF_4, O_2, and He were used to simulate the selective etching of the $Si_{0.7}Ge_{0.3}$/Si. In assessing the results of multilayer structure, the uniformity of etching is an important evaluation index. Figure 3a shows that the profile of the 15-cycle was affected by the gas pressure. Figure 3b shows the range of etching depths for the 15 $Si_{0.7}Ge_{0.3}$ layers under different gas pressures. With the optimization of internal gas uniformity, the etching depths of different layers became more and more consistent, an effect which was also quantified using standard deviation, with the standard deviation value decreasing from 24.98 to 9.21. This random effect may be related to the result of gas diffusion, which leads to the inequal diffusion between the by-products after etching and the etching gas.

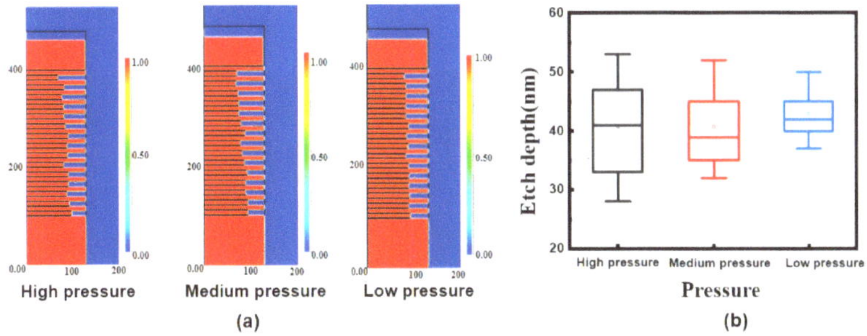

Figure 3. (**a**) Variation of random etching effects in $Si_{0.7}Ge_{0.3}$/Si multilayer structure obtained by PEGASUS 2022 simulation; (**b**) box plot of etching depth of 15 $Si_{0.7}Ge_{0.3}$ layers under different pressure conditions.

In order to truly simulate the ICP etching process, the etching model must include physical (ions) and chemical (radicals) etching processes. In chemical etching, free radicals diffuse to wrap the whole surface area of the sample profile, which is completely isotropic. However, as shown in Figure 4, there is some damage on the Si layer post-etch simulation, and the damage is more severe on its upper surface and is asymmetrical. It is noteworthy that this damage is also mainly concentrated at the nanosheet edges, while the internal profile looks better. Incident ions may be the root cause of this asymmetric damage. The ions are directional after colliding with each other, but most of the damage is concentrated on the upper surface of the Si nanosheets due to the incident angle. In the follow-up experiments, we will conduct an in-depth exploration and analyze the impact of the random effect and the etching damage.

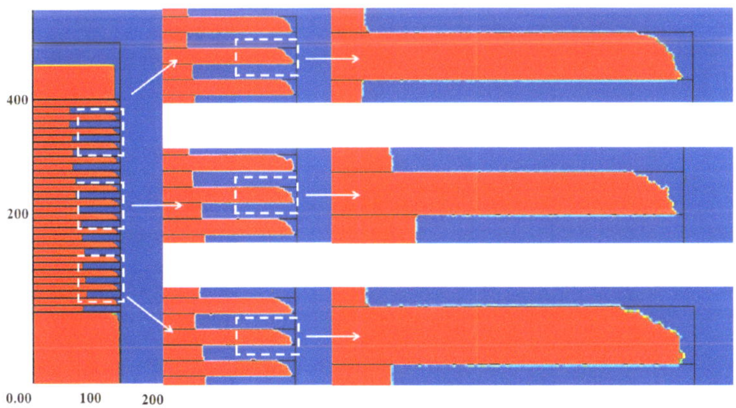

Figure 4. Asymmetry of etching on upper and lower surfaces from PEGASUS 2022 simulation.

3.3. *Pressure Impact on Isotropic Dry Etching*

To verify the simulation results and the impact of pressure on etch performance, we conducted experiments at different gas pressures using ICP. Based on the above-mentioned simulation, we found that the pressure had a great influence on the random effect in the multilayer. Experiments with different pressures were conducted, and the etching results are shown in Figure 5. In the high-pressure range, as shown in Figure 5a, the randomness is relatively large. Even under the same conditions, the etching depth of the $Si_{0.7}Ge_{0.3}$ layer at the same position in the two experiments is not completely repeatable. The etch uniformity is gradually improved as the pressure is reduced, as shown in Figure 5b. As shown in

Figure 5c, the etching profile is quite uniform at low pressure, indicating that the etching depth of each $Si_{0.7}Ge_{0.3}$ layer is consistent, which can be seen in Figure 5d.

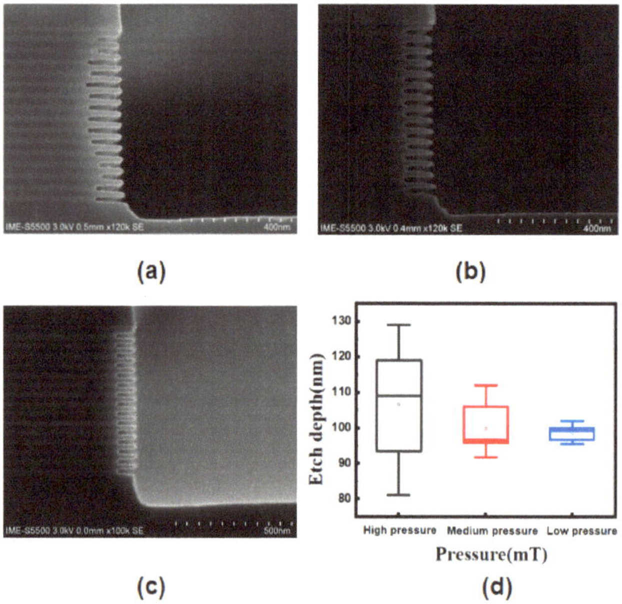

Figure 5. (**a**) Etched profile with high pressure; (**b**) etched profile with medium pressure; (**c**) etched profile with low pressure; and (**d**) box plot of etching depth for 15 $Si_{0.7}Ge_{0.3}$ layers under different pressures.

This phenomenon may be attributed to the result of gas diffusion in the cavity [24]. During the etching process, the lower pressure means that the by-products of the etching are desorbed from the cavity-bottom surface and pumped out at a fast rate and do not significantly hinder the diffusion of reactive gases into the bottom of the cavity. However, under high pressure, the rate at which by-products are drawn out of the cavity slows down, which hinders the diffusion of some etching gases, resulting in random etching effects. This experimental observation is consistent with the previous simulation results, indicating that the uniformity of the gas inside the cavity is affected by the pressure. By adjusting the pressure, the most consistent results after etching of the 15-cycle multilayer structure resulted in a standard deviation σ of 2.06, even better than that of the simulations. In relation to the structure of the seven-levels-stacked nanosheets published in 2020 [21], the result of SiGe selective etching is comparable, and the consistency is greatly improved.

In addition, we set the chamber condition at low pressure ranging from 5 mT to 50 mT and studied how the pressure influenced the $Si_{0.7}Ge_{0.3}$ etching rate and selectivity. As shown in Figure 6, the etching rate corresponding to the right axis is calculated by dividing the measured etching depth by a fixed etching time of 20 s. The plot shows that in the low-pressure regime, the etching rate is faster with increasing pressure. This trend can be explained that, at low pressure, the radical density is lower, resulting in a relatively low etching rate. When the chamber pressure is higher, the plasma density in the reaction chamber increases, more plasma radicals chemically react with the surface of the material, and the etching rate increases. In the pressure range of 5 mT–50 mT, the etch rate increases from 0.90 nm/s to 19.45 nm/s, and it can be seen that the etching rate is very sensitive to pressure changes. According to the fitting curve, the etching rate shows a quadratic relationship with increasing pressure.

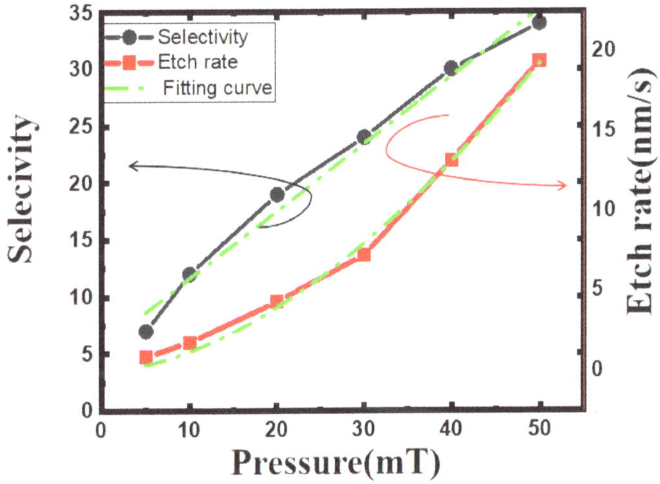

Figure 6. Trend of $Si_{0.7}Ge_{0.3}$ etch rate and selectivity in terms of increasing pressure.

Due to the different etch depths, it is difficult to directly compare the etch profiles under different conditions. We adjusted the etching time under different pressures and fixed the etching depth at around 80 nm. This is because the depth of 80 nm can meet the channel release in all current GAA devices as well as the selective etching of SiGe in 3D-DRAM. Etch selectivity was also calculated and corresponds to the left axis in Figure 6. The selectivity of the $Si_{0.7}Ge_{0.3}$/Si increases continuously with increasing pressure. In the case of lower pressure, the average free path of particles is longer, and the incident energy is higher. Both physical and chemical etching effects are more pronounced. As a result, the etching edge is severely damaged, and the selectivity is lower. As the pressure increases, ion collisions intensify, resulting in a decrease in particle incident energy and thereby reducing the effect of physical etching and improving the selectivity. However, as the by-product is not drawn out of the chamber in time, the edges of the structure become rough. By balancing the selectivity and consistency, the pressure condition was fixed at 20 mT in subsequent experiments.

3.4. Source Power Impact on Isotropic Dry Etching

There are two power sources in the ICP equipment: one is ICP source power and the other is RF bias power. Bias power controls the kinetic energy of particles accelerated to the surface through the electric field, causing the upper-surface damage in the simulation. Its directionality can lead to some degree of anisotropic etching. In order to reduce ion energy to weaken surface physical damage, the RF bias power was turned off. The plasma process was then working in a downstream-like mode.

In order to better understand the role of source power in CF_4/O_2/He plasma, we studied the relationship between the $Si_{0.7}Ge_{0.3}$ etching rate and the power conditions in the etching chamber. As shown in Figure 7, the etching rate corresponding to the right axis is calculated by dividing the measured etching depth by a fixed etching time of 20 s. When the power in the cavity is set in the range of 200–600 W, the etching depth or etching rate increases with the increase in power, and the etching rate increases from 0.94 nm/s to 4.41 nm/s. When the power in the cavity is set in the range of 600–800 W, the etching rate tends to reach saturation and stabilizes at about 4.4 nm/s. Overall, according to the fitting curve, the etching rate has a tendency to be linear and then gradually saturated with increasing power. This is because, as the ICP source power increases, the gas ionization rate increases, and the plasma density increases. Source power generates high-density plasma through inductive coupling, which determines the conversion rate of F radicals and the density of plasma. The increase in reactant ions enhances the chemical reaction and increases the etching rate. But as the source power continues to increase, the plasma density in the reaction chamber tends to saturate, and the chemical reaction between ions and the surface of the material to be etched reaches its peak.

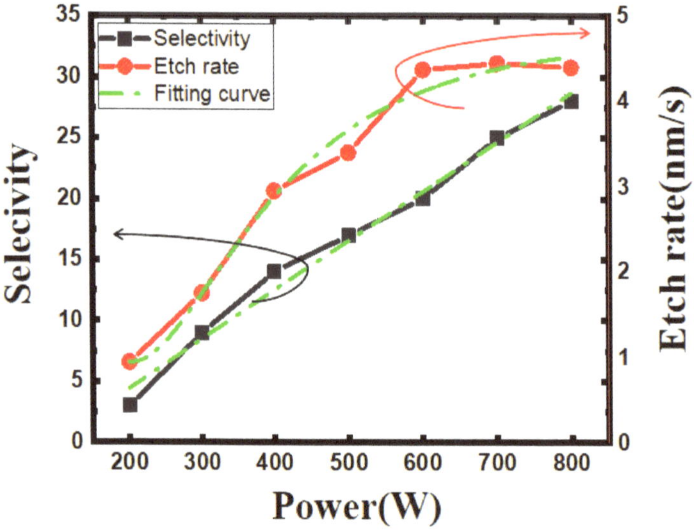

Figure 7. Trend of $Si_{0.7}Ge_{0.3}$ etch rate and selectivity in terms of increasing source power.

In order to explore the influence of different source powers on the $Si_{0.7}Ge_{0.3}$/Si selectivity, we adjusted the etching time under different powers and fixed the etching depth at about 80 nm. Etch selectivity was also calculated and corresponds to the left axis in Figure 7. When the power is in the range of 200–400 W, the lower the power, the greater the Si loss. This is due to the fact that O_2 requires higher power to dissociate compared to CF_4, and the degree of dissociation is lower at low power. And the selectivity of $Si_{0.7}Ge_{0.3}$/Si is very sensitive to the O_2 content, which leads to serious damage to the Si at low power. When the power is in the range of 500–800 W, the selectivity increases with the increase in power, the etching uniformity is better, the etching outline is rectangular, and the angle is relatively sharp. Generally speaking, when the power is in the range of 600–800 W, the etching selectivity reaches a relatively optimized condition.

3.5. Asymmetry Effect of Etching Damage

In order to more accurately characterize the process results in this study, the 15-cycle multilayer of $Si_{0.7}Ge_{0.3}$/Si after selective etching was characterized using TEM. The etching result is shown in Figure 8a. The etching amount of each $Si_{0.7}Ge_{0.3}$ layer is almost equal, with a relatively good uniformity. However, HRTEM results indicate the loss at the upper edge of the nanosheets under conditions of 20 mT and 600 W, resulting in significant silicon damage. It can be inferred from the image that the damage is mainly concentrated on the upper surface of the Si nanosheet edges, which is highly consistent with the phenomenon in the previous simulation.

The root cause of the asymmetric damage is described in Figure 8b. The wafer is grounded and allows positively charged ions to bombard the top surface almost vertically under self-bias. As a result, a portion of particles will bombard the edge of the Si nanosheets due to scattering by collision. During the etching of the $Si_{0.7}Ge_{0.3}$, the upper surface of the nanosheet is more easily damaged, while the lower surface of the nanosheet is significantly less damaged due to the shadowing effect. When we define the angle between the incident direction of the particle and the vertical direction as θ, the incident angle of the particle obeys a normal distribution. In other words, the number of particles with large θ is small. Therefore, the damage on the upper surface is mainly concentrated at the edge, which is an important reason for the selective etching of the $Si_{0.7}Ge_{0.3}$ using ICP being limited.

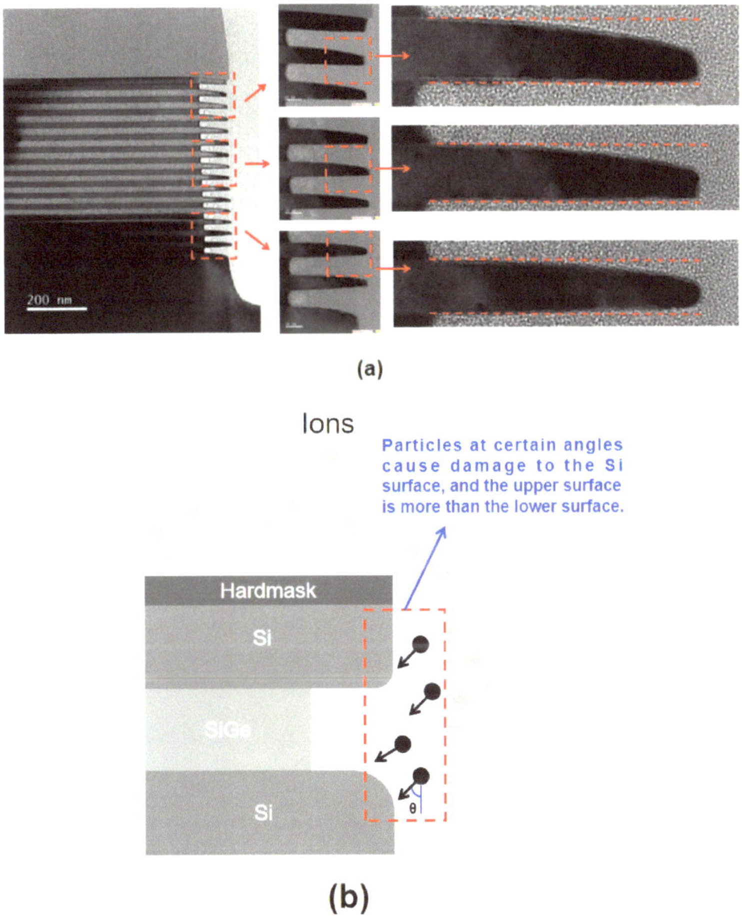

Figure 8. (**a**) TEM images of isotropic etching profile; (**b**) schematic diagram of top surface etch damage.

3.6. Micromorphological Characterization and Material Quality Analyses

Due to the relatively large number of cycled layers, residual by-products or diffusion of elements may occur after etching, so we performed elemental analysis of the etched structure. Figure 9 shows the elemental analysis of a cross section of the sample after etching. The EDS result images show that the boundary of each layer is consistent with the growth of the material, and there is no diffusion and accumulation of Si and Ge elements within the stack after etching. There are no intermediate products containing Ge elements on the surface and no polymers containing C elements that may remain after the use of CF_4 gas. In addition, element C is the loading filler material in the TEM sample, and the oxide layer at the edge of the etched profile is the natural oxide layer formed after the contact of the sample with air.

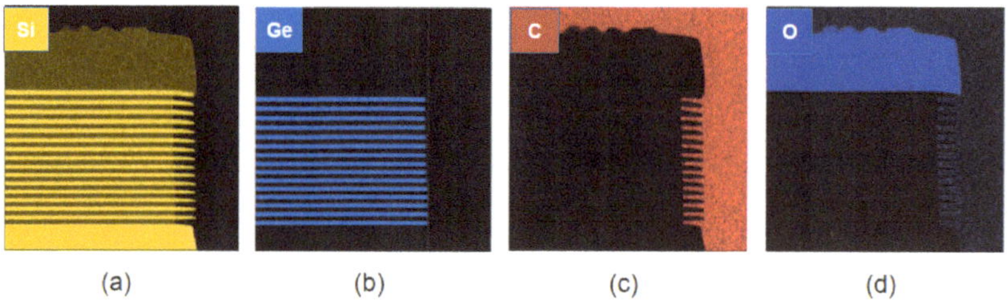

Figure 9. EDS mapping of multilayer structure post-isotropic etching: (**a**) scanning of silicon element of full map; (**b**) scanning of germanium element of full map; (**c**) scanning of carbon element of full map; and (**d**) scanning of oxygen element of full map.

To further analyze the state of the silicon after the etching process, we performed HRXRD characterization of the samples after epitaxy, anisotropic etching, and isotropic etching, respectively. To determine whether our samples were strained or relaxed, an asymmetric scan of (113) facet was required. Figure 10a shows that the SiGe peaks are consistent with the Si peak in the vertical direction, which implies that the epitaxial Si/SiGe multilayer is totally strained. Figure 10b,c show the Si and SiGe peaks after vertical etching and lateral etching, respectively. After vertical etching, the SiGe peaks shift away from the Si peaks, which shows the strain relaxation in the SiGe film [25]. This phenomenon is different from the results for the three-cycle structure in our previous experiments [12], indicating that anisotropic etching is more likely to cause relaxation problems as the number of stacked layers increases. Moreover, this problem still exists after lateral etching. From the experimental results so far, the etching rate of the 15-layer structure is slightly higher than that of the three-layer structure, while the selectivity ratio decreases [12]. We will continue to investigate the influence of the variation of stress on the etching results.

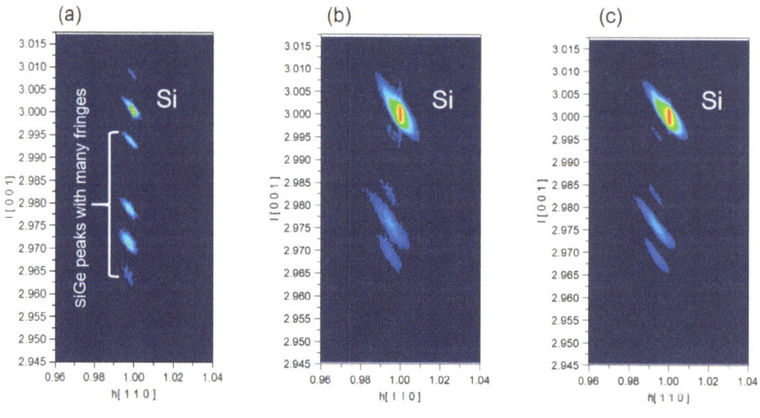

Figure 10. RSMs in the vicinity of the asymmetric (113) Bragg reflection acquired on SiGe/Si multilayer structure: (**a**) unprocessed structure; (**b**) after vertical anisotropic etch and 100:1 DHF wet clean; and (**c**) after SiGe isotropic selectivity etching.

4. Conclusions

This paper investigates the use of a conventional ICP etching system for quasi-isotropic etching of a 15-cycle $Si_{0.7}Ge_{0.3}/Si$ multilayer structure. By conducting simulation and experiments, the study reveals that the selective etching process in the $Si_{0.7}Ge_{0.3}/Si$ multilayer structure causes random effects on the $Si_{0.7}Ge_{0.3}$ layer etch depth and asymmetric damage on the Si surfaces due to radical and ion distribution. Pressure was found to be the main factor for mitigating the random effect, and the standard deviation of the etching depth was reduced by more than 85% to 2.06 by lowering the pressure. For Si nanosheet edge damage, the upper-surface loss was about 2.3 times that of the lower surface; this phenomenon can be explained by bombardment of incident ions accelerated by self-bias. Finally, for a 15-cycle multilayer structure of $Si_{0.7}Ge_{0.3}/Si$ with each layer having a thickness of 20 nm, good etch uniformity and a smooth surface were obtained. The selectivity of etching $Si_{0.7}Ge_{0.3}$ to Si was calculated to be 34 under conditions of 50 mT and 600 W, and an etching rate of 0.90 nm/s~19.45 nm/s was achieved through tuning the process conditions. And it was found that the 15-layer structure was more prone to relaxation in etching than the three-layer structure. In the future, the avoidance of random effects by using low pressure and filtering charged particles during the etching process will offer more application prospects.

Author Contributions: Conceptualization, E.L. and J.L. (Junjie Li); methodology, E.L. and J.L. (Junjie Li); software, E.L., R.C., H.S. and P.L.; validation, E.L., H.L. and J.L. (Junjie Li); formal analysis, E.L.; investigation, E.L., C.Y. and Y.L.; resources, J.L. (Junjie Li), R.C., G.W., Z.K. and Q.Z.; data curation, E.L.; writing—original draft preparation, E.L.; writing—review and editing, E.L., J.L. (Junjie Li), R.C., G.W. and H.S.; visualization, E.L. and C.Z. (Chenchen Zhang); supervision, J.L. (Jun Luo), N.Z., R.C., G.W., C.Z. (Chao Zhao) and H.S.; project administration, J.G., T.Y., H.Y., J.L. (Junfeng Li), J.L. (Jun Luo) and W.W.; funding acquisition, J.L. (Junjie Li) and G.W. All authors have read and agreed to the published version of the manuscript.

Funding: This work was financially supported by the Strategic Priority Research Program of the Chinese Academy of Sciences (Grant No. XDA0330401); the Chinese Academy of Sciences Supporting Technology Talent Project (Grant No. E2YR01 × 001); the Beijing Superstring Academy of Memory Technology (Grant No. E2DF06 × 003); Development of dry release function for gate-all-around (GAA) nanosheet devices (Grant No. E2SH01X); and the Innovation Program for Quantum Science and Technology (Grant No. 2021ZD0302300).

Data Availability Statement: The data can be obtained from the authors upon request.

Conflicts of Interest: The authors declare no conflict of interest.

References

1. Moore, G.E. Cramming more components onto integrated circuits. *IEEE Solid-State Circuits Newsl.* **1965**, *11*, 33–35. [CrossRef]
2. Li, J.; Li, Y.; Zhou, N.; Xiong, W.; Wang, G.; Zhang, Q.; Du, A.; Gao, J.; Kong, Z.; Lin, H.; et al. Study of silicon nitride inner spacer formation in process of gate-all-around nano-transistors. *Nanomaterials* **2020**, *10*, 793. [CrossRef] [PubMed]
3. Sun, X.; Wang, D.; Qian, L.; Liu, T.; Yang, J.; Chen, K.; Wang, L.; Huang, Z.; Xu, M.; Wang, C.; et al. A Novel Si Nanosheet Channel Release Process for the Fabrication of Gate-All-Around Transistors and Its Mechanism Investigation. *Nanomaterials* **2023**, *13*, 504. [CrossRef] [PubMed]
4. Loubet, N.; Hook, T.; Montanini, P.; Yeung, C.-W.; Kanakasabapathy, S.; Guillom, M.; Yamashita, T.; Zhang, J.; Miao, X.; Wang, J.; et al. Stacked nanosheet gate-all-around transistor to enable scaling beyond FinFET. In Proceedings of the 2017 Symposium on VLSI Technology, Kyoto, Japan, 5–8 June 2017; IEEE: Piscataway, NJ, USA, 2017; pp. T230–T231.
5. Wang, Y.; Chen, Z.; Wang, C.; Keller, N.; Antonelli, G.A.; Liu, Z.; Ribaudo, T.; Grynko, R. 3D NAND Oxide/Nitride Tier Stack Thickness Measurements with Infrared Metrology. In Proceedings of the 2023 34th Annual SEMI Advanced Semiconductor Manufacturing Conference (ASMC), Saratoga Springs, NY, USA, 1–4 May 2023; IEEE: Piscataway, NJ, USA, 2023; pp. 1–8.
6. Han, J.; Park, S.; Jeong, M.; Lee, K.S.; Kim, K.N.; Kim, H.J.; Shin, J.C.; Park, S.M.; Shin, S.H.; Park, S.W.; et al. Ongoing Evolution of DRAM Scaling via Third Dimension—Vertically Stacked DRAM. In Proceedings of the 2023 Symposium on VLSI Technology, Kyoto, Japan; IEEE: Piscataway, NJ, USA, 2023.
7. Chu, C.L.; Wu, K.; Luo, G.L.; Chen, B.Y.; Chen, S.H.; Wu, W.F.; Yeh, W.K. Stacked Ge-nanosheet GAAFETs fabricated by Ge/Si multilayer epitaxy. *IEEE Electron Device Lett.* **2018**, *39*, 1133–1136. [CrossRef]
8. Borel, S.; Arvet, C.; Bilde, J.; Caubet, V.; Louis, D. Control of Selectivity between SiGe and Si in isotropic etching processes. *Jpn. J. Appl. Phys.* **2004**, *43*, 3964. [CrossRef]
9. Tsai, Y.H.; Wang, M. Fundamental study on the selective etching of SiGe and Si in ClF_3 gas for nanosheet gate-all-around transistor manufacturing: A first principle study. *J. Vac. Sci. Technol. B* **2022**, *40*, 013201. [CrossRef]
10. Zhang, Q.; Gu, J.; Xu, R.; Cao, L.; Li, J.; Wu, Z.; Wang, G.; Yao, J.; Zhang, Z.; Xiang, J.; et al. Optimization of structure and electrical characteristics for four-layer vertically-stacked horizontal gate-all-around Si nanosheets devices. *Nanomaterials* **2021**, *11*, 646. [CrossRef] [PubMed]
11. Mertens, H.; Ritzenthaler, R.; Hikavyy, A.; Kim, M.S.; Tao, Z.; Wostyn, K.; Chew, S.A.; De Keersgieter, A.; Mannaert, G.; Rosseel, E.; et al. Gate-all-around MOSFETs based on vertically stacked horizontal Si nanowires in a replacement metal gate process on

bulk Si substrates. In Proceedings of the 2016 IEEE Symposium on VLSI Technology, Honolulu, HI, USA, 14–16 June 2016; IEEE: Piscataway, NJ, USA, 2016; pp. 1–2.
12. Li, J.; Wang, W.; Li, Y.; Zhou, N.; Wang, G.; Kong, Z.; Fu, J.; Yin, X.; Li, C.; Wang, X.; et al. Study of selective isotropic etching $Si_{1-x}Ge_x$ in process of nanowire transistors. *J. Mater. Sci. Mater. Electron.* **2020**, *31*, 134–143. [CrossRef]
13. Xue, Z.Y.; Wei, X.; Liu, L.J.; Chen, D.; Zhang, B.; Zhang, M.; Wang, X. Etch characteristics of Si_{1-x} Gex films in HNO_3: H_2O: HF. *Sci. China Technol. Sci.* **2011**, *54*, 2802–2807. [CrossRef]
14. Baraissov, Z.; Pacco, A.; Koneti, S.; Bisht, G.; Panciera, F.; Holsteyns, F.; Mirsaidov, U. Selective wet etching of silicon germanium in composite vertical nanowires. *ACS Appl. Mater. Interfaces* **2019**, *11*, 36839–36846. [CrossRef] [PubMed]
15. Hartmann, J.M.; Destefanis, V.; Rabillé, G.; Monfray, S. HCl selective etching of SiGe versus Si in stacks grown on (1 1 0). *Semicond. Sci. Technol.* **2010**, *25*, 105009. [CrossRef]
16. Oehrlein, G.S.; Kroesen, G.M.W.; De Frésart, E.; Zhang, Y.; Bestwick, T.D. Studies of the reactive ion etching of SiGe alloys. *J. Vac. Sci. Technol. A Vac. Surf. Film.* **1991**, *9*, 768–774. [CrossRef]
17. Kumar, A.; Lee, W.H.; Wang, Y.L. Optimizing the Isotropic Etching Nature and Etch Profile of Si, Ge and $Si_{0.8}Ge_{0.2}$ by Controlling CF4 Atmosphere with Ar and O_2 Additives in ICP. *IEEE Trans. Semicond. Manuf.* **2021**, *34*, 177–184. [CrossRef]
18. Barnola, S.; Vizioz, C.; Vulliet, N.; Dupré, C.; Ernst, T.; Gautier, P.; Arvet, C.; Guillaumot, B.; Bernard, E.; Pauliac-Vaujeour, S.; et al. Dry etch challenges in gate all around devices for sub 32 nm applications. *ECS Trans.* **2008**, *16*, 923–934. [CrossRef]
19. Zhao, Y.; Iwase, T.; Satake, M.; Hamamura, H. Formation mechanism of a rounded SiGe-etch-front in an isotropic dry SiGe etch process for gate-all-around (GAA)-FETs. In Proceedings of the 2021 5th IEEE Electron Devices Technology & Manufacturing Conference (EDTM), Chengdu, China, 8–11 April 2021; IEEE: Piscataway, NJ, USA, 2021; pp. 1–3.
20. Loubet, N.; Kal, S.; Alix, C.; Pancharatnam, S.; Zhou, H.; Durfee, C.; Belyansky, M.; Haller, N.; Watanabe, K.; Devarajan, T.; et al. A novel dry selective etch of SiGe for the enablement of high performance logic stacked gate-all-around nanosheet devices. In Proceedings of the 2019 IEEE International Electron Devices Meeting (IEDM), San Francisco, CA, USA, 7–11 December 2019; IEEE: Piscataway, NJ, USA, 2019; pp. 11.4.1–11.4.4.
21. Liu, Y.C.; Tu, C.T.; Tsai, C.E.; Huang, B.W.; Cheng, C.Y.; Chueh, S.J.; Chen, J.Y.; Liu, C.W. Highly stacked GeSi nanosheets and nanowires by low-temperature epitaxy and wet etching. *IEEE Trans. Electron Devices* **2021**, *68*, 6599–6604. [CrossRef]
22. Barraud, S.; Previtali, B.; Vizioz, C.; Hartmann, J.M.; Sturm, J.; Lassarre, J.; Perrot, C.; Rodriguez, P.; Loup, V.; Magalhaes-Lucas, A.; et al. 7-levels-stacked nanosheet GAA transistors for high performance computing. In Proceedings of the 2020 IEEE Symposium on VLSI Technology, Honolulu, HI, USA, 16–19 June 2020; IEEE: Piscataway, NJ, USA, 2020; pp. 1–2.
23. Radamson, H.H.; Zhu, H.; Wu, Z.; He, X.; Lin, H.; Liu, J.; Xiang, J.; Kong, Z.; Xiong, W.; Li, J.; et al. State of the art and future perspectives in advanced CMOS technology. *Nanomaterials* **2020**, *10*, 1555. [CrossRef] [PubMed]
24. Cunningham, R.E.; Williams, R.J.J. *Diffusion in Gases and Porous Media*; Plenum Press: New York, NY, USA, 1980.
25. Brunner, K. Si/ge nanostructures. *Rep. Prog. Phys.* **2001**, *65*, 27. [CrossRef]

nanomaterials MDPI

Article

Insight into over Repair of Hot Carrier Degradation by GIDL Current in Si p-FinFETs Using Ultra-Fast Measurement Technique

Hao Chang [1,2], Guilei Wang [3], Hong Yang [1,2,*], Qianqian Liu [1], Longda Zhou [4], Zhigang Ji [4], Ruixi Yu [1,2], Zhenhua Wu [1,2], Huaxiang Yin [1,2], Anyan Du [1,2], Junfeng Li [1,2], Jun Luo [1,2], Chao Zhao [3] and Wenwu Wang [1,*]

1 Key Laboratory of Microelectronics Devices & Integrated Technology, Institute of Microelectronics, Chinese Academy of Sciences, Beijing 100029, China; changhao@ime.ac.cn (H.C.)
2 Microelectronics Institute, University of Chinese Academy of Sciences, Beijing 100049, China
3 Process Integration, Beijing Superstring Academy of Memory Technology, Beijing 100176, China
4 National Key Laboratory of Science and Technology on Micro/Nano Fabrication, Shanghai Jiaotong University, Shanghai 200240, China
* Correspondence: yanghong@ime.ac.cn (H.Y.); wangwenwu@ime.ac.cn (W.W.)

Abstract: In this article, an experimental study on the gate-induced drain leakage (GIDL) current repairing worst hot carrier degradation (HCD) in Si p-FinFETs is investigated with the aid of an ultra-fast measurement (UFM) technique (~30 μs). It is found that increasing GIDL bias from 3 V to 4 V achieves a 114.7% V_T recovery ratio from HCD. This over-repair phenomenon of HCD by UFM GIDL is deeply discussed through oxide trap behaviors. When the applied gate-to-drain GIDL bias reaches 4 V, a significant electron trapping and interface trap generation of the fresh device with GIDL repair is observed, which greatly contributes to the approximate 114.7% over-repair V_T ratio of the device under worst HCD stress (−2.0 V, 200 s). Based on the TCAD simulation results, the increase in the vertical electric field on the surface of the channel oxide layer is the direct cause of an extraordinary electron trapping effect accompanied by the over-repair phenomenon. Under a high positive electric field, a part of channel electrons is captured by oxide traps in the gate dielectric, leading to further V_T recovery. Through the discharge-based multi-pulse (DMP) technique, the energy distribution of oxide traps after GIDL recovery is obtained. It is found that over-repair results in a 34% increment in oxide traps around the conduction energy band (E_c) of silicon, which corresponds to a higher stabilized V_T shift under multi-cycle HCD-GIDL tests. The results provide a trap-based understanding of the transistor repairing technique, which could provide guidance for the reliable long-term operation of ICs.

Keywords: reliability; hot carrier degradation (HCD); Si p-FinFETs; gate-induced drain leakage (GIDL); recovery; oxide trap generation; energy distribution

1. Introduction

DRAM (dynamic random access memory) is one of the core components of electronic equipment, and it has become increasingly important in terms of the development of the information society [1]. For higher memory density, the downscaling of transistor size has become an inevitable trend. Fin Field-Effect Transistors (FinFETs) are recognized as one of the most promising structures for Future DRAM Peripheral Circuits and have been proposed for low-power and high-performance applications beyond 22-nm technology nodes [2,3].

With continuous channel length scaling, hot carrier degradation (HCD) has emerged as a major reliability issue in FinFETs [4]. From the trap-based research in HCD, both interface and oxide traps contribute to the overall degradation [5]. A modified compact model and trap spatial distribution investigations facilitate the accurate characterization of HCD [6–8]. To characterize the trap generation during HCD stress, a discharge-based multi-pulse technique (DMP) was introduced, which is accessible to oxide traps within and beyond the bandgap [9,10].

To ensure the reliable long-term operation of the transistor in ICs, controlling gate oxide quality is one of the most critical challenges. Several methods have been developed to recover the hot-carrier-induced damage. From the wafer-level view, the forming gas annealing (FGA) process utilizes high-pressure hydrogen or deuterium to passivate the dangling bonds at the interface and thus suppress the interface trap generation (ΔN_{IT}) [11,12]. The recovery effects exhibit a positive correlation with annealing temperature, which could be attributed to the thermally activated interface traps discharging [13,14]. From the transistor-level view, an electrothermal annealing (ETA) method shows its feasibility in curing degraded gate oxide. The early ETA method achieves thermal annealing by external micro-heaters but may cause over-heating in metal interconnections due to heat diffusion [15]. Another implementation utilizes the Joule heat inherently generated in the device by the flowing current as the heat source, such as Punch-through and GIDL currents [16,17]. The new methods show superior annealing selectivity and can cure the target transistor, which has experienced severe HCD [18]. This active recovery capability enables ETA to demonstrate better applicability in actual circuits [19]. However, previous ETA methods mainly focus on SOI and GAAFET structures, which themselves have poor heat dissipation performance. In a recent study, the GIDL repairing method showed its feasibility in Bulk FinFETs, which is motivated by field-assisted discharging [20].

However, the trap behaviors during the GIDL repairing process for HCD still remain to be discussed. Another issue that needs to be discussed is the underlying mechanism of the observed V_T over-recovery phenomenon. With the development of the ultra-fast measurement technique (UFM), the recovery effect during I_D-V_G measurement is effectively suppressed, further improving the accuracy in HCD characterization [21,22]. In this article, the UFM with a microsecond (~30 μs) delay is used for device characterization. The recovery behaviors of p-FinFETs with 100 nm gate length are investigated. Additionally, the trap behaviors during GIDL repair are discussed with the aid of DMP [23,24]. With the aid of technology computer-aided design (TCAD) tools, the mechanism of over-repairing is explained from the perspective of electric-field simulation. The results provide experimental evidence of the GIDL recovery-related traps and their energy locations, which could provide further understanding of FET recovery techniques.

The remainder of this manuscript is organized as follows. Section 2 elucidates the device under test, the measurement methods, and the TCAD simulation setup. The test results with discussions are shown in Section 3. The conclusions are summarized in Section 4.

2. Materials and Methods

2.1. Device Fabrication

The replacement metal gate (RMG) Si bulk p-FinFETs are fabricated using a fully-gate-last process. The equivalent oxide thickness (EOT) is 0.92 nm. The major steps for gate stack formation are shown in Figure 1a: (1) dummy poly-Si/SiO_2 gate removal; (2) the growth of the 0.8 nm interface layer (IL) of SiO_2 through chemical oxidation of O_3; (3) the atom layer deposition (ALD) of 1.7 nm HfO_2 as a high-k layer; (4) 450 °C post-deposition annealing (PDA); and (5) the deposition of a multi-layer gate stack including ALD Titanium Nitride (TiN)/ ALD Tantalum Nitride (TaN) /CVD Titanium Nitride (TiN)/ ALD Tungsten (W). Figure 1b is the transmission electron microscope (TEM) image of FinFET across the channel direction. P-FinFETs with 100 nm gate length are used for electrical measurements.

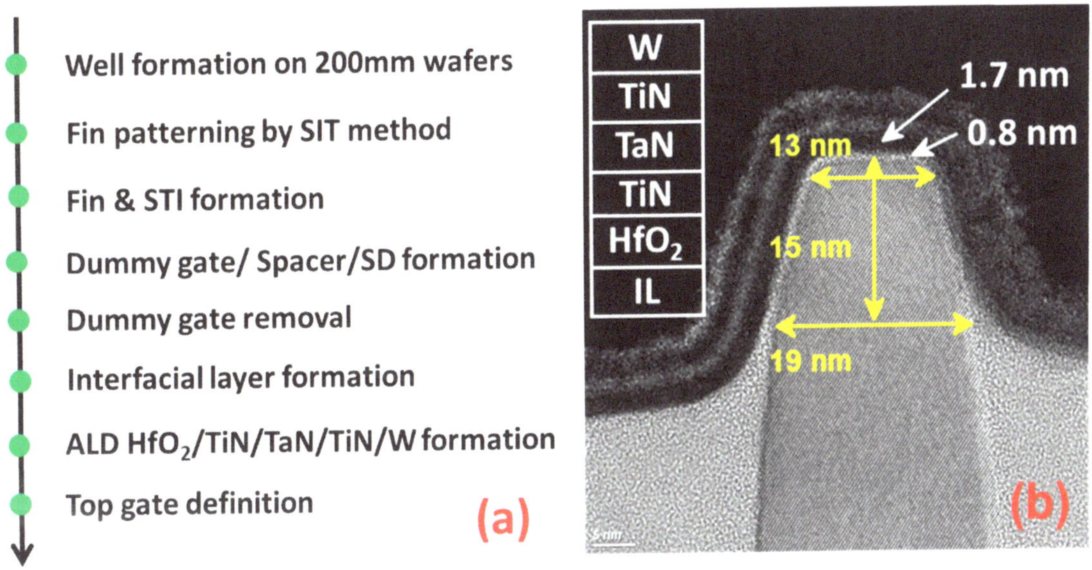

Figure 1. (**a**) The process of p-FinFET with replacement metal gate (RMG). (**b**) TEM image of p-FinFET across channel direction.

2.2. Electrical Measurements

Devices are stressed under the worst HCD condition ($V_{G,STR}$ = $V_{D,STR}$) [6]; then, a GIDL voltage is applied to the stressed devices using a synchronized pulse of specified gate ($V_{G,GIDL}$) and drain GIDL biases ($V_{D,GIDL}$). In this article, the Keysight B1530 semiconductor analyzer is utilized to perform UFM of I_D-V_G characteristics using pulse-IV measurements with a duration of 30 μs [22]. Time evolutions of threshold voltage shift (ΔV_T) are obtained in measure-stress-measure (MSM) mode [25]. V_T is extracted through the constant current method with the target linear drain current ($I_{D,LIN}$) of 100 nA × W/L [26]; here, W and L are the gate width and length, respectively.

The DMP experiments are performed to investigate the energy distribution of generated oxide traps during GIDL repair. Figure 2 shows the DMP test procedure used in this work. After 200 s HCD stress, 1 ks GIDL bias is applied to repair the aged device. Afterward, the repaired device is applied with the same HCD stress again; then, V_G is sequentially decreased to multiple gate discharge voltage ($V_{G,DIS}$) levels, while V_D remains at the stress bias. Each discharge period lasts for only 1 s; then, a pulse-IV is performed to extract ΔV_T. The overdrive voltage (V_{OV}) is calculated by V_{OV} = $V_{G,DIS}$—V_T; then, the ΔV_T ~ V_{OV} relationship is obtained to extract the energy distribution of oxide traps. All of the test results are averaged by a group of three devices at 125 °C.

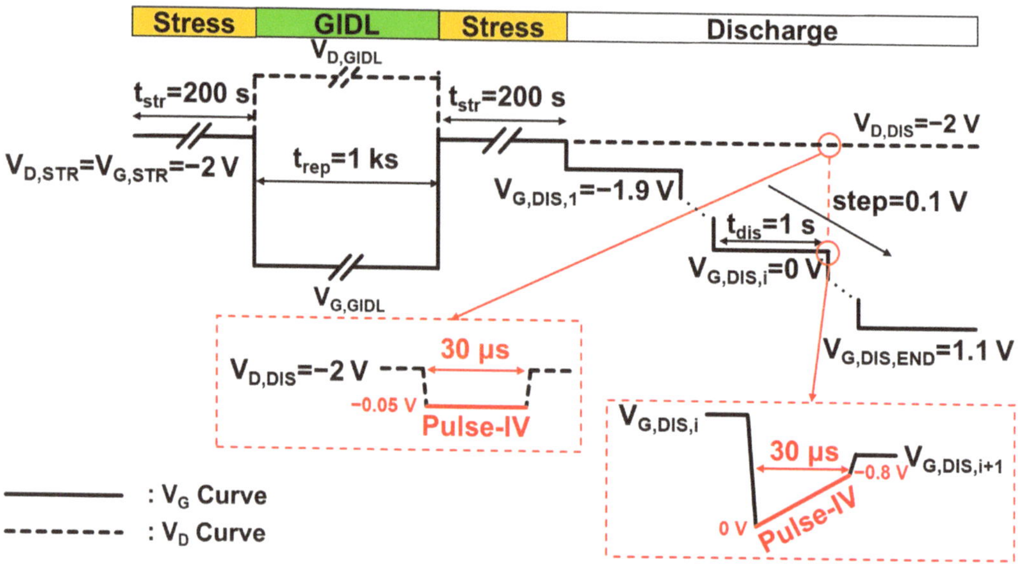

Figure 2. V_D and V_G test waveforms for the DMP test. Solid curve: V_G test waveform. Dashed curve: V_D test waveform. Red curve: Pulse-IV measurement.

2.3. *TCAD Simulation Setup*

To bring a further physical explanation, Sentaurus TCAD tools are employed to solve the electric field distribution of p-FinFETs under GIDL repair. The 3-D simulation structure of p-FinFETs with the same Fin shape in TEM is shown in Figure 3a. As shown in Figure 3b, the simulated I_D-V_G curve with GIDL is in good agreement with measured data within 1 m V_T. The key simulation parameters, such as work function, stress, and the S/D distribution resistance, are concluded in Table 1. Here, the Nonlocal-Path model is used as a band-to-band physical model to accurately match GIDL characteristics in TCAD simulation [27].

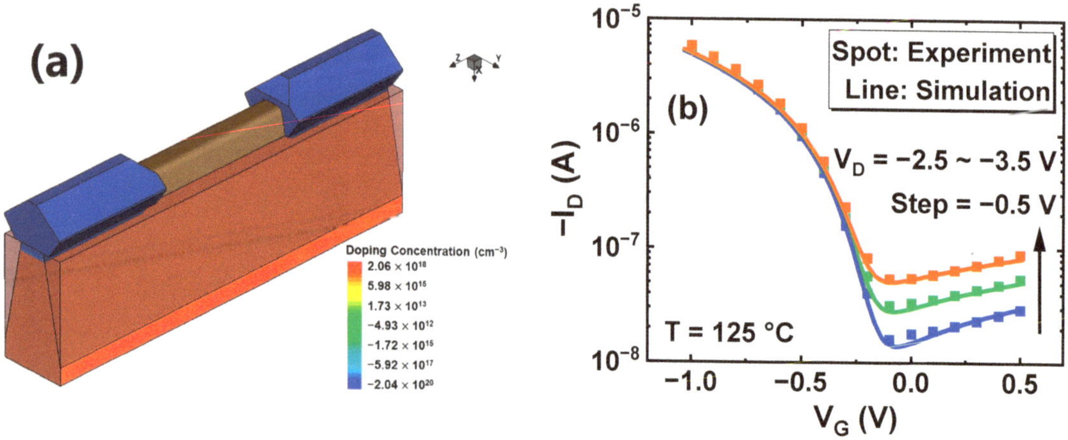

Figure 3. (**a**) 3-D simulation structures of Bulk FinFET with 100 nm gate length. (**b**) TCAD calibration of Bulk FinFET I_D-V_G and GIDL characteristics with experimental data. Orange curve: V_D = −3.5 V. Green curve: V_D = −3.0 V. Blue curve: V_D = −2.5 V.

Table 1. Key device parameters used for TCAD simulation.

Parameter	Bulk FinFET
Channel length (L_G)	100 nm
Fin height	15 nm
Fin top width	13 nm
Fin bottom width	19 nm
Effective oxide thickness (EOT)	0.92 nm
Source/drain doping	$2 \times 10^{20}/cm^3$
Gate work function	4.97 eV
Source/drain distribution resistance	$4.4 \times 10^{-8}\ \Omega cm^2$

3. Results and Discussion

3.1. Repairing HCD by UFM GIDL

Figure 4 shows that p-FinFETs with 100 nm gate length are subjected to a −2 V HCD stress for 200 s, resulting in a V_T degradation of approximately 170 mV and a shift of 8 mV/dec in subthreshold swing (SS). Subsequently, the UFM GIDL repairing process is implemented for 1 ks with different repairing biases. Under moderate GIDL bias of V_{GD} = 3 V ($V_{G,GIDL}$ = 0.5 V, $V_{D,GIDL}$ = −2.5 V), the V_T shift is reduced to 60 mV, and the SS shift is negligible, corresponding to a 62.7% recovery ratio, as shown in Figure 4a. In Figure 4b, under a high GIDL bias of V_{GD} = 4 V ($V_{G,GIDL}$ = 1.5 V, $V_{D,GIDL}$ = −2.5 V), the V_T recovery ratio is increased to 114.7%, which indicates that the degraded device is over-repaired by GIDL repair. However, this improvement comes at the cost of degradation in SS of 9 mV/dec. Considering that SS degradation reflects the generation rate of interface traps [28], excessive gate-to-drain electric field may result in extraordinary interface damage to the device [29].

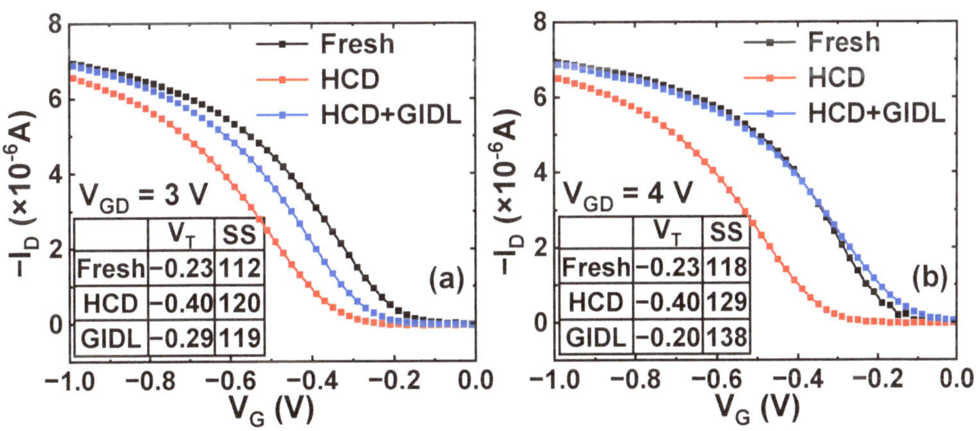

Figure 4. Measured pulse-IV characteristics of p-FinFETs at the initial state, after 200 s HCD, and after 1 ks GIDL repair for (**a**) V_{GD} = 3 V: $V_{G,GIDL}$ = 0.5 V, $V_{D,GIDL}$ = −2.5 V and (**b**) V_{GD} = 4 V: $V_{G,GIDL}$ = 1.5 V, $V_{D,GIDL}$ = −2.5 V. HCD stress: $V_{G,STR}$ = $V_{D,STR}$ = −2 V.

3.2. Physical Explainations of GIDL Repairing Mechanism

To explain the over-repair phenomenon that occurs at high GIDL biases, a group of fresh devices is subjected to 1 ks GIDL biases, as shown in Figure 5. When a moderate GIDL bias ($V_{G,GIDL}$ = 0.5 V, $V_{D,GIDL}$ = −2.5 V) is applied, the recovery of V_T is relatively low, measuring approximately 27 mV. However, under high GIDL biases (by either increasing $V_{G,GIDL}$ or $V_{D,GIDL}$), the recovery of V_T exceeds 100 mV. It can be seen that at high gate-to-drain electric fields, a large number of electrons are injected into the gate oxide, which greatly contributes to V_T recovery [30]. Moreover, after applying a high GIDL bias, a degradation of approximately 12 mV/dec in SS can be observed. However, no SS degradation is observed at a moderate GIDL bias.

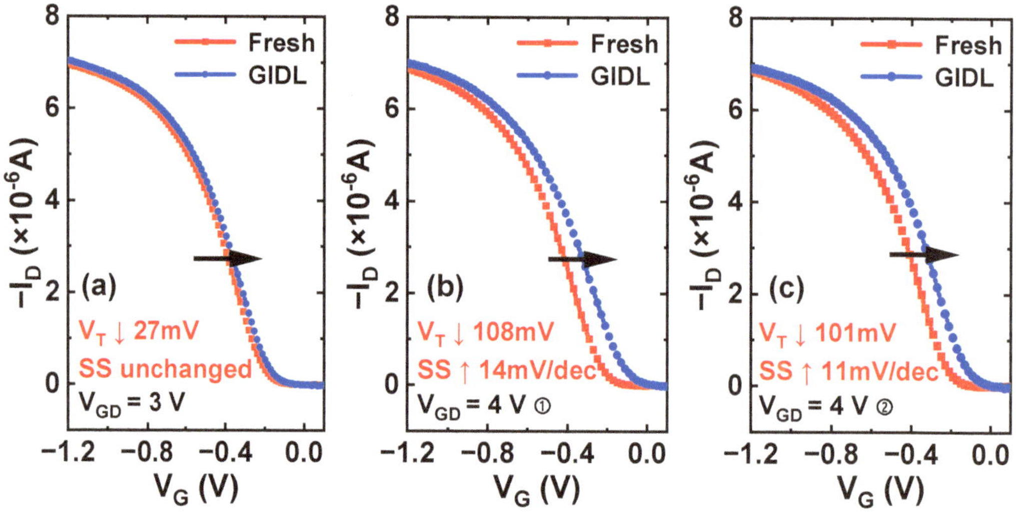

Figure 5. Pulse-IV curves of fresh p-FinFETs before and after 1 ks GIDL biases for (**a**) V_{GD} = 3 V: $V_{G,GIDL}$ = 0.5 V, $V_{D,GIDL}$ = −2.5 V; (**b**) V_{GD} = 4 V①: $V_{G,GIDL}$ = 1.5 V, $V_{D,GIDL}$ = −2.5 V; and (**c**) V_{GD} = 4 V②: $V_{G,GIDL}$ = 0.5 V, $V_{D,GIDL}$ = −3.5 V.

Furthermore, a physical explanation is given with the aid of TCAD simulation. The electric field distributions of the device under different GIDL biases are simulated. Cutting vertically the center of gate along the channel direction, the 2-D view of the electric field distributions is shown in Figure 6. As V_{GD} increases from 3 V to 4 V, the channel electric field changes significantly at two positions: one near the drain region, and the other along the channel surface. The 1-D electric field, cut along the surface of oxide layer in 2-D view in Figure 6, is shown in Figure 7a. It is shown that the electric field peaks near the drain region. Then, this peak electric field and the channel center electric field are extracted in Figure 7b and named as $E_{x=50}$ and $E_{x=0}$ according to their positional coordinates.

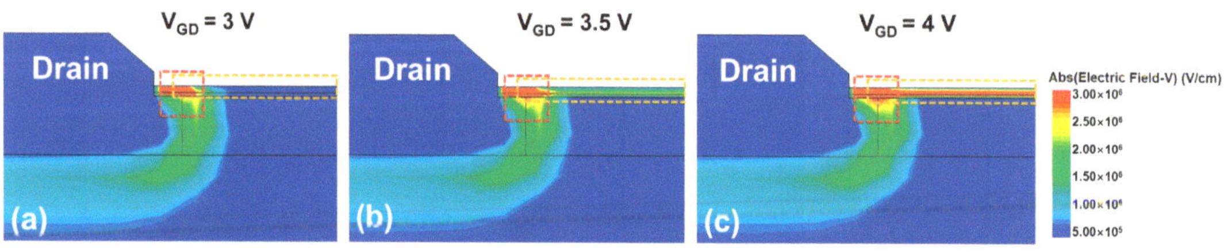

Figure 6. The 2-D view of electric field distributions under GIDL repairing process of (**a**) V_{GD} = 3.0 V, (**b**) V_{GD} = 3.5 V, and (**c**) V_{GD} = 4.0 V.

As V_{GD} increases from 3 V to 4 V, $E_{x=50}$ and $E_{x=0}$ increase by 1.26 and 6.89 times, respectively. As shown in Figure 8, the influences of these two electric fields will be discussed separately. During HCD stress, the channel hot holes near the drain region are captured by the oxide traps in gate dielectric, resulting in the V_T degradation. When a GIDL bias is applied, the strong positive electric field near the drain region promotes the trapped holes to be released, which results in V_T recovery [31]. The higher $E_{x=50}$ at V_{GD} = 4 V corresponds to the higher V_T recovery ratio (114.7%) in Figure 4. While a high V_{GD} is also like applying a PBTI stress to the device, part of electrons in the channel are captured by the HfO_2 traps, which also results in the recovery in V_T [32]. The higher $E_{x=0}$ at V_{GD} = 4 V is the main cause of over 100 mV V_T recovery in Figure 5. More discharged holes and trapped electrons directly lead to the over-recovery phenomenon observed at high GIDL biases.

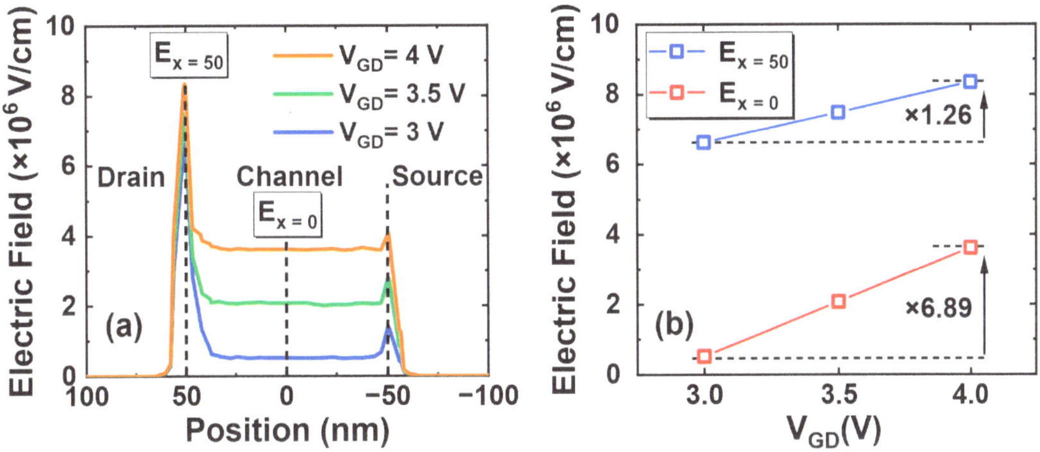

Figure 7. (**a**) Electric field distribution along the surface of the oxide layer in 2-D view under different GIDL biases (V_{GD} = 3.0 V/3.5 V/4.0 V). (**b**) Extracted $E_{x=50}$ and $E_{x=0}$ values at different GIDL biases.

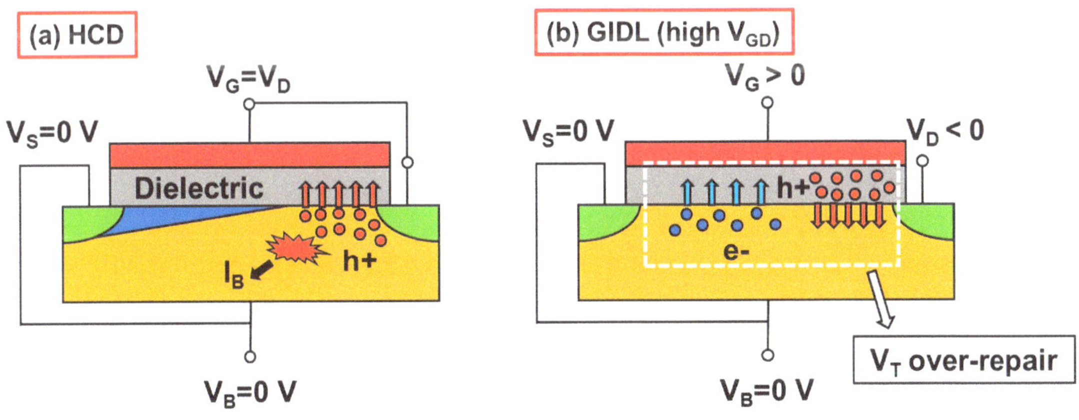

Figure 8. Schematic diagrams during (**a**) HCD and (**b**) GIDL repair with high V_{GD}. H+ and e- are short for holes and electrons in the channel, respectively.

3.3. Oxide Trap Behaviors during GIDL Repairing

To further analyze the trap behaviors during the GIDL repairing process, DMP experiments are carried out. A group of fresh p-FinFETs is subjected to −2 V HCD stress for 200 s, followed by a 1 ks GIDL repairing process at different GIDL biases. The repaired devices are again applied with the same HCD stress and then discharged following the waveform in Figure 2. Additionally, $\Delta V_T \sim V_{OV}$ relationships under different GIDL biases are obtained in Figure 9a. Then, ΔV_T is converted into the effective trap density (ΔN_T) following the equation $\Delta N_T = |\Delta V_T| \times C_{ox} / q$ [23]. V_{OV} is converted into the corresponding energy level E_f relative to E_v, i.e., E_f—E_v. Additionally, the $\Delta N_T \sim (E_f$—$E_v)$ relationships under different GIDL biases are obtained in Figure 9b. Finally, the energy density of ΔN_T (ΔD_T) is obtained by differentiating ΔN_T against E_f—E_v, and the results are shown in Figure 10a. The detailed processing flow can be seen in the previous research [10]. For comparison, a group of devices under the same HCD stress are directly discharged without GIDL repair. Additionally, the energy density of pre-existing traps (ΔD_{HT}) is extracted through the multi-DMP method separately, which is represented by a black curve [24].

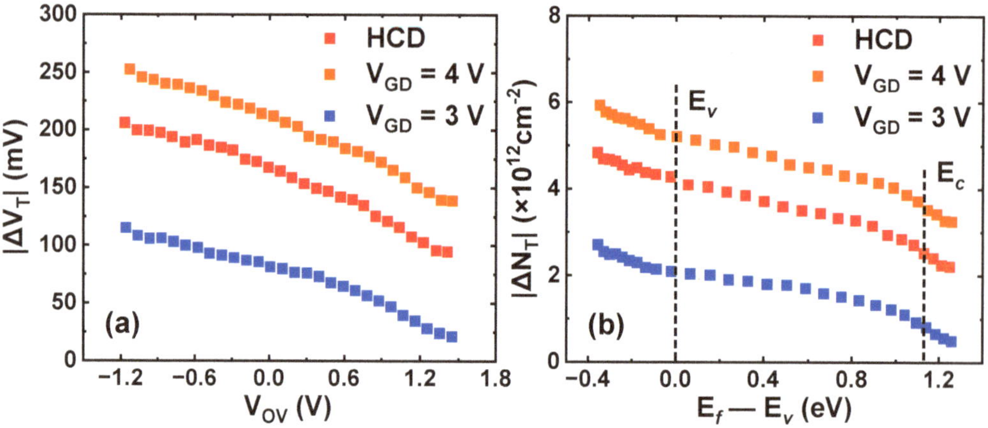

Figure 9. Extracted (**a**) $\Delta V_T \sim V_{OV}$ and (**b**) $\Delta N_T \sim (E_f—E_v)$ relationships after applying a new round of HCD stress to recovered devices by different GIDL biases. V_{GD} = 3 V: $V_{G,GIDL}$ = 0.5 V, $V_{D,GIDL}$ = −2.5 V. V_{GD} = 4 V: $V_{G,GIDL}$ = 1.5 V, and $V_{D,GIDL}$ = −2.5 V.

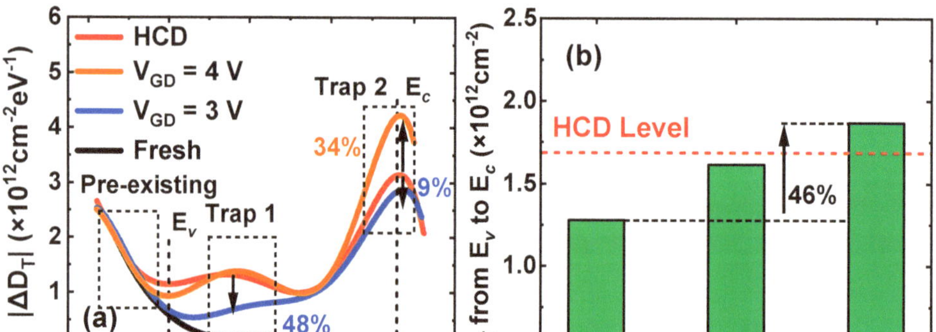

Figure 10. (**a**) Extracted energy distribution of ΔD_T in the new round of HCD stress after repairing by different GIDL biases. "Fresh" is ΔD_{HT} extracted by multi-DMP method separately. (**b**) Extracted ΔN_T from E_v to E_c in the new round of HCD stress. Fixed $V_{D,GIDL}$ = −2.5 V; $V_{G,GIDL}$ varies from 0.5 V to 1.5 V.

As can be seen from Figure 10a, the overall degradation of HCD can be divided into three components: pre-existing traps below E_v, and two types of generated oxide traps in the bandgap. Generated trap 1 is at 0.2 eV below the mid-gap of silicon, and trap 2 is located around E_c of silicon. The pre-existing traps are mainly induced by the fabrication process and cannot be repaired by GIDL bias. While generated traps show clear dependence with applied GIDL biases. At a moderate GIDL bias of V_{GD} = 3 V, when a new round of HCD stress is applied, the ΔD_T of both trap 1 and trap 2 are reduced by 48% and 9% compared to that under HCD, respectively. While at a high GIDL bias of V_{GD} = 4 V, the ΔD_T of trap 2 is 34% higher than that under HCD. In addition, the energy position of trap 1 tends to approach E_c after repair, while the position of trap 2 remains unchanged. To make a clear comparison, ΔD_T was integrated between E_v and E_c to extract ΔN_T, as shown in Figure 10b. As V_{GD} increases from 3 V to 4 V, ΔN_T is increased by 46%. It can be seen that, after applying a moderate GIDL bias, the oxide trap density is effectively reduced in the next round of HCD stress. However, at high GIDL biases, although more generated oxide traps are discharged after repair, they are re-charged in the next round of HCD stress, which results in the further generation of oxide trap 2.

To verify the effectiveness of different GIDL biases on long-term HCD recovery, multiple cycles of HCD and GIDL are applied, and the results are shown in Figure 11a. At a high GIDL bias of V_{GD} = 4 V, the repaired device exhibits a lower V_T (below the pre-stress value) but shows more severe degradation in the next round of HCD stress. As the HCD/GIDL cycle progressed, the degraded ΔV_T and recovered ΔV_T in each cycle gradually tended to be the same. As can be seen from Figure 11b, the stabilized ΔV_T at high GIDL bias is almost 99.4% higher than that at moderate GIDL bias, which corresponds to the higher ΔN_T at high GIDL bias in Figure 10b. The consistent trends between ΔN_T and ΔV_T illustrate that the cyclic charge–discharge behavior of oxide traps results in the dynamic equilibrium between V_T degradation and recovery. To ensure the reliable operation of p-FinFETs under long-term HCD stress, moderate GIDL bias is recommended to ensure the minimum ΔV_T under cyclic stress.

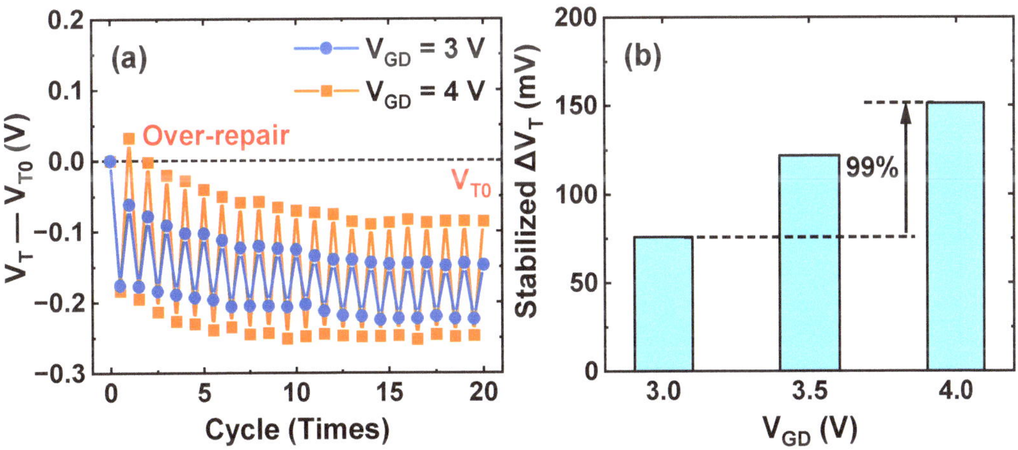

Figure 11. (**a**) Evolutions of V_T variation (V_T—V_{T0}) during 20 cycles of HCD/GIDL at different GIDL biases. Each cycle contains 200 s HCD stress ($V_{G,STR} = V_{D,STR} = -2$ V) followed by 1 ks GIDL repair. (**b**) Stabilized ΔV_T during 20 cycles of HCD/GIDL at different GIDL biases. Fixed $V_{D,GIDL} = -2.5$ V; $V_{G,GIDL}$ varies from 0.5 V to 1.5 V.

Finally, the energy level diagrams of the two generated traps are given in Figure 12. As described in the As-grown-generation (AG) model [33], two types of oxide traps with different discharge characteristics have been observed in the silicon energy band. One of them captures a hole without energy level changing, which corresponds to trap 2 around E_c. Due to the shallow energy level of trap 2, it will be charged first during the second round of HCD stress, which corresponds to the higher ΔD_T peak observed in Figure 10a. While trap 1 is a type of energy alternating defects (EADs), the energy position shifts towards E_c after capturing a hole. Trap 1 is located at a relatively deep level (0.2 eV below mid-gap of silicon), which will be fully charged only after a complete discharge. Therefore, under a moderate GIDL bias of V_{GD} = 3 V, the ΔD_T of trap 1 shows a 48% decrease compared to

HCD. For the convenience of reading, all abbreviations and notations frequently used in this article can be seen in Table 2.

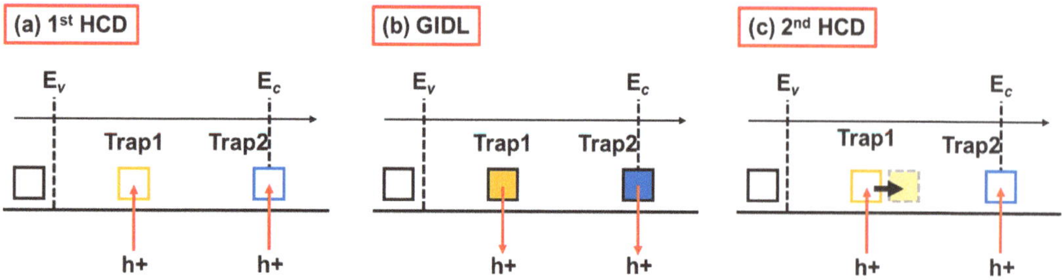

Figure 12. Energy level diagrams during (**a**) the first round of HCD stress, (**b**) GIDL repairing process, and (**c**) the second round of HCD stress.

Table 2. List of abbreviations and notations frequently used in this article.

Abbreviations/Notations	Meaning
HCD	Hot carrier degradation
GIDL	Gate-induced drain leakage
FinFETs	Fin field-effect transistors
UFM	Ultra-fast measurement
DMP	Discharge-based multi-pulse technique
SS	Subthreshold swing
PBTI	Positive bias temperature instability
TCAD	Technology computer-aided design
ΔV_T	Threshold voltage shift
$V_{G,STR}$/ $V_{D,STR}$	Gate/drain stress voltage
$V_{G,GIDL}$/ $V_{D,GIDL}$	Gate/drain GIDL repair bias
V_{GD}	Gate-to-drain bias
V_{OV}	Overdrive voltage
$E_{x=50}$	Channel electric field peak near the drain region (at x = 50)
$E_{x=0}$	Channel center electric field (at x = 0)
ΔN_T	Effective trap density
ΔD_T	Energy density of ΔN_T
E_v/ E_c	Valance/conduction energy band of silicon

4. Conclusions

In this article, the GIDL repairing process is carried out with the UFM technique, and trap generations during GIDL repair are experimentally investigated. At a high GIDL bias of V_{GD} = 4 V, the V_T recovery ratio reaches 114.7%. With a 6.89 times increase in the channel electric field, more PBTI components are introduced at high V_{GD}, which are responsible for additional electron trapping and interface trap generation. At a moderate GIDL bias of V_{GD} = 3 V, the effective density of generated oxide trap 1 (at 0.2 eV below mid-gap of silicon) and trap 2 (around E_c of silicon) are reduced by 48% and 9% in the next round of HCD stress. However, a high GIDL bias will lead to 34% further generation of trap 2. Furthermore, two generated traps show different charge–discharge properties, which corresponds to two types of oxide traps described in AG model. After multiple cycles of HCD/GIDL tests, the degraded and recovered V_T reaches the same. The dynamic equilibrium between V_T degradation and recovery can be attributed to the cyclic charge–discharge behaviors of oxide traps. To ensure long-term HCD reliability, a moderate GIDL bias is recommended.

Author Contributions: Writing—original draft preparation, H.C.; review and editing, H.Y. (Hong Yang) and G.W.; device fabrication, Q.L. and G.W.; methodology, H.C., L.Z., Z.J. and Z.W.; investigation, H.C. and R.Y.; project administration, H.Y. (Huaxiang Yin), A.D., J.L. (Junfeng Li) and C.Z.; supervision, H.Y. (Hong Yang) and J.L. (Jun Luo) and W.W. All authors have read and agreed to the published version of the manuscript.

Funding: This work was financially supported by the National Natural Science Foundation of China (Grant No. 92064002).

Institutional Review Board Statement: Not applicable.

Informed Consent Statement: Not applicable.

Data Availability Statement: The data may be obtained from the authors upon request.

Conflicts of Interest: The authors declare no conflict of interest.

References

1. An, H.D.; Lee, S.H.; Park, J.; Min, S.R.; Kim, G.U.; Yoon, Y.J.; Seo, J.H.; Cho, M.S.; Jang, J.; Bae, J.-H.; et al. Design of a Capacitorless DRAM Based on a Polycrystalline-Silicon Dual-Gate MOSFET with a Fin-Shaped Structure. *Nanomaterials* **2022**, *12*, 3526. [CrossRef]
2. Capogreco, E.; Arimura, H.; Ritzenthaler, R.; Brus, S.; Oniki, Y.; Dupuy, E.; Sebaai, F.; Radisic, D.; Chan, B.T.; Zhou, D.; et al. FinFETs with Thermally Stable RMG Gate Stack for Future DRAM Peripheral Circuits. In Proceedings of the 2022 International Electron Devices Meeting (IEDM), San Francisco, CA, USA, 3–7 December 2022; pp. 26.22.21–26.22.24.
3. Gupta, A.; Gupta, C.; Vega, R.A.; Hook, T.B.; Dixit, A. Reliability Modeling and Analysis of Hot-Carrier Degradation in Multiple-Fin SOI n-Channel FinFETs with Self-Heating. *IEEE Trans. Electron Devices* **2019**, *66*, 2075–2080. [CrossRef]
4. Chung, E.A.; Nam, K.J.; Nakanishi, T.; Park, S.; Yang, H.; Kauerauf, T.; Jiao, G.; Kim, D.W.; Hwang, K.H.; Kim, H.; et al. Investigation of hot carrier degradation in bulk FinFET. In Proceedings of the 2017 IEEE International Reliability Physics Symposium (IRPS), Monterey, CA, USA, 2–6 April 2017; pp. XT-6.1–XT-6.4.
5. Yu, Z.; Zhang, J.; Wang, R.; Guo, S.; Liu, C.; Huang, R. New insights into the hot carrier degradation (HCD) in FinFET: New observations, unified compact model, and impacts on circuit reliability. In Proceedings of the 2017 IEEE International Electron Devices Meeting (IEDM), San Francisco, CA, USA, 2–6 December 2017; pp. 7.2.1–7.2.4.
6. Yu, Z.; Sun, Z.; Wang, R.; Zhang, J.; Huang, R. Hot Carrier Degradation-Induced Dynamic Variability in FinFETs: Experiments and Modeling. *IEEE Trans. Electron Devices* **2020**, *67*, 1517–1522. [CrossRef]
7. Yu, Z.; Zhang, Z.; Sun, Z.; Wang, R.; Huang, R. On the Trap Locations in Bulk FinFETs After Hot Carrier Degradation (HCD). *IEEE Trans. Electron Devices* **2020**, *67*, 3005–3009. [CrossRef]
8. Ding, Y.; Yan, C.; Qu, Y.; Zhao, Y. Re-Examination of Hot Carrier Degradation Mechanism in Ultra-Scaled nFinFETs. *IEEE Electron Device Lett.* **2022**, *43*, 1802–1805. [CrossRef]
9. Gao, R.; Ji, Z.; Zhang, J.F.; Zhang, W.D.; Hatta, S.F.W.M.; Niblock, J.; Bachmayr, P.; Stauffer, L.; Wright, K.; Greer, S. A Discharge-Based Pulse Technique for Probing the Energy Distribution of Positive Charges in Gate Dielectric. *IEEE Trans. Semicond. Manuf.* **2015**, *28*, 221–226. [CrossRef]
10. Chang, H.; Zhou, L.; Yang, H.; Ji, Z.; Liu, Q.; Simoen, E.; Yin, H.; Wang, W. Comparative Study on the Energy Distribution of Defects under HCD and NBTI in Short Channel p-FinFETs. In Proceedings of the 2021 IEEE International Reliability Physics Symposium (IRPS), Monterey, CA, USA, 21–25 March 2021; pp. 1–5.
11. Pobegen, G.; Tyaginov, S.; Nelhiebel, M.; Grasser, T. Observation of Normally Distributed Energies for Interface Trap Recovery After Hot-Carrier Degradation. *IEEE Electron Device Lett.* **2013**, *34*, 939–941. [CrossRef]
12. Park, H.; Choi, R.; Lee, B.H.; Hwang, H. Observation of Normally Distributed Energies for Interface Semiconductor Field Effect Transistors with High-k Gate Dielectric by Using High Pressure Deuterium Post Metallization Annealing. *Jpn. J. Appl. Phys.* **2007**, *46*, L786. [CrossRef]
13. Jong, M.J.; Salm, C.; Schmitz, J. Towards understanding recovery of hot-carrier induced degradation. *Microelectron. Reliab.* **2018**, *88*, 147–151. [CrossRef]
14. Son, D.; Kim, G.J.; Kim, J.; Lee, N.; Kim, K.; Pae, S. Effect of High Temperature on Recovery of Hot Carrier Degradation of scaled nMOSFETs in DRAM. In Proceedings of the 2021 IEEE International Reliability Physics Symposium (IRPS), Monterey, CA, USA, 21–25 March 2021; pp. 1–4.
15. Han, J.W.; Kebaili, M.; Meyyappan, M. System on Microheater for On-Chip Annealing of Defects Generated by Hot-Carrier Injection, Bias Temperature Instability, and Ionizing Radiation. *IEEE Electron Device Lett.* **2016**, *37*, 1543–1546. [CrossRef]
16. Park, J.Y.; Hur, J.; Choi, Y.K. Demonstration of a Curable Nanowire FinFET Using Punchthrough Current to Repair Hot-Carrier Damage. *IEEE Electron Device Lett.* **2018**, *39*, 180–183. [CrossRef]
17. Park, J.Y.; Yun, D.H.; Choi, Y.K. Curing of Hot-Carrier Induced Damage by Gate-Induced Drain Leakage Current in Gate-All-Around FETs. *IEEE Electron Device Lett.* **2019**, *40*, 1909–1912. [CrossRef]

18. Park, J.Y.; Lee, G.B.; Choi, Y.K. A Comparative Study of the Curing Effects of Local and Global Thermal Annealing on a FinFET. *IEEE J. Electron Devices Soc.* **2019**, *7*, 954–958. [CrossRef]
19. Park, J.Y.; Moon, D.I.; Lee, G.B.; Choi, Y.K. Curing of Aged Gate Dielectric by the Self-Heating Effect in MOSFETs. *IEEE Trans. Electron Devices* **2020**, *67*, 777–788. [CrossRef]
20. Chang, H.; Liu, Q.; Yang, H.; Zhou, L.; Ji, Z.; Tang, B.; Zhang, Q.; Yin, H.; Du, A.; Li, J.; et al. Effectiveness of Repairing Hot Carrier Degradation in Si p-FinFETs Using Gate Induced Drain Leakage. *IEEE Electron Device Lett.* **2023**, *44*, 372–375. [CrossRef]
21. Qu, Y.; Lin, X.; Li, J.; Cheng, R.; Yu, X.; Zheng, Z.; Lu, J.; Chen, B.; Zhao, Y. Ultra fast (<1 ns) electrical characterization of self-heating effect and its impact on hot carrier injection in 14nm FinFETs. In Proceedings of the 2017 IEEE International Electron Devices Meeting (IEDM), San Francisco, CA, USA, 2–6 December 2017; pp. 39.32.31–39.32.34.
22. Cheng, R.; Yu, X.; Shen, L.; Yin, L.; Zhang, Y.; Zheng, Z.; Chen, B.; Liu, X.; Zhao, Y. Ultrafast pulse characterization of hot carrier injection effects on ballistic carrier transport for sub-100 nm MOSFETs. In Proceedings of the 2017 IEEE International Reliability Physics Symposium (IRPS), Monterey, CA, USA, 2–6 April 2017; pp. 3E-3.1–3E-3.6.
23. Hatta, S.W.M.; Ji, Z.; Zhang, J.F.; Duan, M.; Zhang, W.D.; Soin, N.; Kaczer, B.; Gendt, S.D.; Groeseneken, G. Energy Distribution of Positive Charges in Gate Dielectric: Probing Technique and Impacts of Different Defects. *IEEE Trans. Electron Devices* **2013**, *60*, 1745–1753. [CrossRef]
24. Ji, Z.; Zhang, X.; Franco, J.; Gao, R.; Duan, M.; Zhang, J.F.; Zhang, W.D.; Kaczer, B.; Alian, A.; Linten, D.; et al. An Investigation on Border Traps in III–V MOSFETs With an $In_{0.53}Ga_{0.47}As$ Channel. *IEEE Trans. Electron Devices* **2015**, *62*, 3633–3639. [CrossRef]
25. Shen, C.; Li, M.F.; Foo, C.E.; Yang, T.; Huang, D.M.; Yap, A.; Samudra, G.S.; Yeo, Y.C. Characterization and Physical Origin of Fast Vth Transient in NBTI of pMOSFETs with SiON Dielectric. In Proceedings of the 2006 International Electron Devices Meeting, San Francisco, CA, USA, 11–13 December 2006; pp. 1–4.
26. Bucher, M.; Makris, N.; Chevas, L. Generalized Constant Current Method for Determining MOSFET Threshold Voltage. *IEEE Trans. Electron Devices* **2020**, *67*, 4559–4562. [CrossRef]
27. Liu, C.; Yu, W.; Zhang, B.; Wang, X.; Zhao, Q.T. Investigation of band-to-band tunneling parameters in SiGe by using MOSFET GIDL current analysis. In Proceedings of the 2014 12th IEEE International Conference on Solid-State and Integrated Circuit Technology (ICSICT), Guilin, China, 28–31 October 2014; pp. 1–3.
28. Wang, M.; Liu, Z.; Yamashita, T.; Stathis, J.H.; Chen, C.Y. Separation of interface states and electron trapping for hot carrier degradation in ultra-scaled replacement metal gate n-FinFET. In Proceedings of the 2015 IEEE International Reliability Physics Symposium, Monterey, CA, USA, 19–23 April 2015; pp. 4A.5.1–4A.5.7.
29. Varghese, D.; Kufluoglu, H.; Reddy, V.; Shichijo, H.; Mosher, D.; Krishnan, S.; Alam, M.A. Off-State Degradation in Drain-Extended NMOS Transistors: Interface Damage and Correlation to Dielectric Breakdown. *IEEE Trans. Electron Devices* **2007**, *54*, 2669–2678. [CrossRef]
30. Tkachev, Y. Field-induced generation of electron traps in the tunnel oxide of flash memory cells. In Proceedings of the 2015 IEEE International Integrated Reliability Workshop (IIRW), South Lake Tahoe, CA, USA, 11–15 October 2015; pp. 99–102.
31. Sun, Z.; Yu, Z.; Zhang, Z.; Zhang, J.; Wang, R.; Lu, P.; Huang, R. Investigation on the Lateral Trap Distributions in Nanoscale MOSFETs During Hot Carrier Stress. *IEEE Electron Device Lett.* **2019**, *40*, 490–493. [CrossRef]
32. Tang, H.; Xu, H.; Chen, L.; Zhu, H.; Sun, Q. Anomalous PBTI Effects in N-Type Super Junction under High Gate Voltage Stress. *Electronics* **2022**, *11*, 1362. [CrossRef]
33. Gao, R.; Ji, Z.; Zhang, J.F.; Marsland, J.; Zhang, W.D. As-grown-Generation Model for Positive Bias Temperature Instability. *IEEE Trans. Electron Devices* **2018**, *65*, 3662–3668. [CrossRef]

nanomaterials

MDPI

Article

High-Quality Recrystallization of Amorphous Silicon on Si (100) Induced via Laser Annealing at the Nanoscale

Zhuo Chen [1,2], Huilong Zhu [1,*], Guilei Wang [3], Qi Wang [1], Zhongrui Xiao [1,2], Yongkui Zhang [1], Jinbiao Liu [1,2], Shunshun Lu [1], Yong Du [1], Jiahan Yu [1,2], Wenjuan Xiong [1], Zhenzhen Kong [1,2], Anyan Du [1], Zijin Yan [1,2] and Yantong Zheng [1,2]

1 Key Laboratory of Microelectronics Devices & Integrated Technology, Institute of Microelectronics, Chinese Academy of Sciences, Beijing 100029, China; chenzhuo@ime.ac.cn (Z.C.); wangqi@ime.ac.cn (Q.W.); xiaozhongrui@ime.ac.cn (Z.X.); zhangyongkui@ime.ac.cn (Y.Z.); liujinbiao@ime.ac.cn (J.L.); lushunshun@ime.ac.cn (S.L.); duyong@ime.ac.cn (Y.D.); yujiahan@ime.ac.cn (J.Y.); xiongwenjuan@ime.ac.cn (W.X.); kongzhenzhen@ime.ac.cn (Z.K.); duanyan@ime.ac.cn (A.D.); yanzijin@ime.ac.cn (Z.Y.); zhengyantong@ime.ac.cn (Y.Z.)

2 Microelectronics Institute, University of Chinese Academy of Sciences, Beijing 100049, China

3 Process Integration, Beijing Superstring Academy of Memory Technology, Beijing 100176, China; guilei.wang@bjsamt.org.cn

* Correspondence: zhuhuilong@ime.ac.cn

Abstract: At sub-3 nm nodes, the scaling of lateral devices represented by a fin field-effect transistor (FinFET) and gate-all-around field effect transistors (GAAFET) faces increasing technical challenges. At the same time, the development of vertical devices in the three-dimensional direction has excellent potential for scaling. However, existing vertical devices face two technical challenges: "self-alignment of gate and channel" and "precise gate length control". A recrystallization-based vertical C-shaped-channel nanosheet field effect transistor (RC-VCNFET) was proposed, and related process modules were developed. The vertical nanosheet with an "exposed top" structure was successfully fabricated. Moreover, through physical characterization methods such as scanning electron microscopy (SEM), atomic force microscopy (AFM), conductive atomic force microscopy (C-AFM) and transmission electron microscopy (TEM), the influencing factors of the crystal structure of the vertical nanosheet were analyzed. This lays the foundation for fabricating high-performance and low-cost RC-VCNFETs devices in the future.

Keywords: vertical nanosheet; laser annealing; recrystallization; Si cap

1. Introduction

With the evolution of Moore's law, it becomes more and more difficult to scale down transistor size [1]. The 2 nm technology nodes Samsung, Intel, and TSMC will all have the architecture of gate-all-around field effect transistors (GAAFETs) [2–5]. The dynamic random-access memory (DRAM) roadmap of the International Roadmap for Devices and Systems (IRDS) 2020 report proposes that the cell transistor structure of DRAM will shift from one of the current mainstream Saddle Fin to the vertical channel transistor (VCT) [6–11]. In logic applications, IBM and Samsung jointly proposed vertical-transport FET (VTFET), which achieved a 40 nm contacted gate pitch (CGP) under excellent gate control, which is significantly lower than the 45 nm CGP of the TSMC 3 nm fin field-effect transistor (FinFET) technology node [12–14]. This proves that the vertical device has great potential for future device footprint scaling. However, the patterning of the vertical transistor channel mainly relies on advanced lithography and etching, which are accompanied by large process fluctuations. Implementing the self-alignment technology in lateral devices for gates and device channels is also challenging. As a result, there are large fluctuations in the vertical device's performance.

Based on the advanced Si/SiGe/Si epitaxy and SiGe selective etching process, self-aligned vertical sandwich FET (VSAFET) and vertical C-shaped-channel nanosheet FET (VCNFET) devices were proposed successively [15–19]. In this way, gate and source/drain self-alignment was realized. At the same time, a device with a strong gate control ability and a large driving current was prepared. However, the above devices all use expensive epitaxial processes. At the same time, introducing SiGe epitaxy at the front-end-of-line (FEOL) of the process flow caused the problem of Ge contamination. Controlling the contamination of Ge during mass production significantly reduces the versatility of related equipment, making it difficult to process other Ge-free devices, and thus increasing the cost of mass production.

This paper proposes a recrystallization-based vertical C-shaped-channel nanosheet field effect transistor (RC-VCNFET) and process integration method. The entire process of the device did not involve the Ge element, avoiding the use of expensive epitaxial processes. This method utilizes the principle of laser annealing and recrystallization of the a-Si. A high-quality recrystallized C-shaped Si channel was fabricated at the nanometer scale, by optimizing the laser annealing energy and other process parameters. Through physical characterization methods, it was found that the lattice structure of Si in the channel is very close to a single-crystal structure. This result can provide a foundation for the future fabrication of low-cost, high-mobility vertical channel devices.

2. Materials and Methods

Figure 1a–h illustrates the main process flow for forming the recrystallized vertical nanosheet. First, 80 nm SiN was deposited on a silicon substrate via plasma-enhanced chemical vapor deposition (PECVD, AMAT Producer S PECVD, Applied Materials, Santa Clara, CA, USA), and 180 nm a-Si was deposited at 580 °C via rapid thermal chemical vapor deposition (RTCVD, Centura, Applied Materials, Santa Clara, CA, USA). Next, 10 nm SiO_2, 300 nm a-Si and 300 nm SiO_2 stacks (abbreviated as OSO stacks) were sequentially deposited via PECVD. Moreover, the OSO hard mask (OSO HM) structure in Figure 1a was formed via lithography and etching. Then, silicon oxide sidewalls were formed sequentially, depositing an oxide and anisotropic etching oxide, as shown in Figure 1b. Then, using the silicon oxide as a hard mask, the a-Si/SiN/c-Si stacks were etched through an anisotropic RIE process to form the structure shown in Figure 1c. Then, SiN was etched isotropically by 160 °C H_3PO_4 to form a C-shaped-cavity structure, as shown in Figure 1d. Next, a 20 nm thick Si cap was grown via RTCVD. When growing the Si cap, a diluted buffered oxide etchant (dBOE) was used to remove the natural oxide layer on the c-Si surface in the C-shaped-cavity. After the Si cap growth was completed, the Si cap on the OSO HM and oxide spacer was removed, forming the structure in Figure 1e via RIE. Then a high-aspect-ratio-process(HARP) oxide was deposited via PECVD, and the wafer was polished via a chemical mechanical planarization (CMP, FRX200, Ebara, Tokyo, Japan) process until the mandrel in the OSO HM was exposed. Moreover, the a-Si was removed using a high-selectivity TMAH wet etch. Next, 10 nm silicon oxide CESL was etched via RIE. Next, using silicon oxide as a mask, the inner a-Si/SiN/c-Si stacks were etched via RIE to form the structure shown in Figure 1f. Next, the remaining SiN in the device cavity was removed using 160 °C H_3PO_4. Then, the HARP oxide was deposited again and CMP was performed on it so that the height of the silicon oxide was about 20 nm from the top of the a-Si. Then, the diluted hydrofluoric acid solution (dHF) was used for the oxide recess process so that the surface of the HARP oxide was lowered to the position shown in Figure 1g. At this point, about half of the top a-Si of the RC-VCNFET was exposed. Subsequently, four groups of Nd:YLF pulsed lasers with different energy densities were used to irradiate the wafer's surface (the laser annealing equipment was developed by the Institute of Microelectronics, Chinese Academy of Sciences, the laser's wavelength is 527 nm, the pulse width is 200 ns, and the frequency is 200 Hz). At this time, the a-Si began to recrystallize under laser light irradiation. Finally, we continued to etch the HARP oxide through the STI recess process to release the RC channel. In the next experiment, if

RC-VCNFET devices need to be fabricated, process steps such as the gate stacks formation and subsequent BEOL should be carried out.

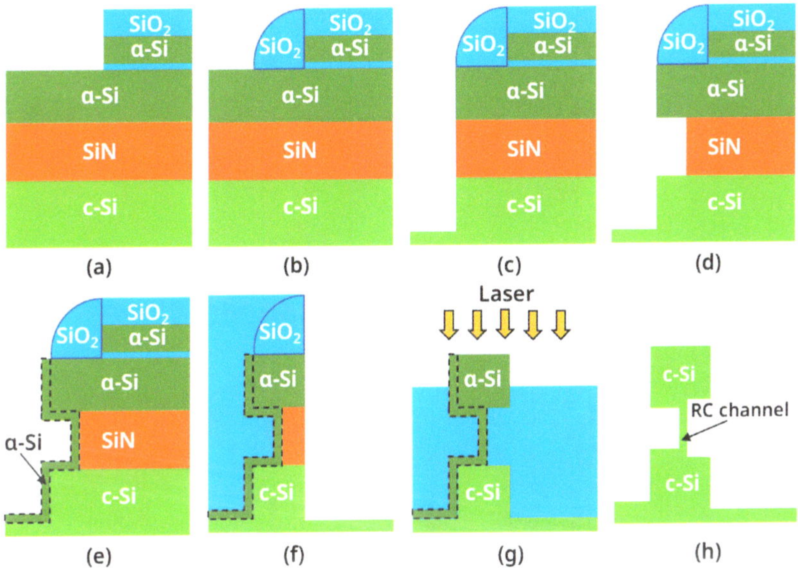

Figure 1. Schematic diagram of the key process steps for the RC-VCNFETs: (**a**) OSO HM formation; (**b**) oxide spacer formation; (**c**) a-Si/SiN/c-Si etching; (**d**) SiN etching; (**e**) Si cap depsition; (**f**) inner a-Si/SiN/c-Si etching; (**g**) STI recess and laser annealing; (**h**) RC channel release.

In addition, we also conducted a short loop of the Si cap growth module and the flow chart is shown in Figure 2a. We have reported the experiment of Group A, in another work. In that experiment, the blank wafers used all had high-energy boron ion implantation. Therefore, we conducted new experiments on Group B, and the blank wafers in this group of experiments did not have a p-well. The following are the experimental steps. Firstly, a pre-clean step was performed on two groups of silicon wafers, one without dBOE etching and the other with 60 s dBOE etching. These two groups of wafers were respectively named wafer(B-1) and wafer(B-2). Subsequently, these wafers were immediately loaded into the chamber of RTCVD, thereby reducing the formation of the natural oxide layer on the surface of the wafer. Next, a 40 nm-thick Si cap was grown on the surface of the wafer at 580 °C.

Scanning electron microscopy (SEM, S-5500, Hitachi, Tokyo, Japan) was used to observe the topography of the surface and cross-section of the sample, thereby measuring the film thickness and etching depth. Atomic force microscopy (AFM, Dimension Icon, Bruker, Karlsruhe, Germany) was used to evaluate the film surface's roughness. Conductive atomic force microscopy (C-AFM, Dimension Icon, Bruker, Karlsruhe, Germany) was used to characterize the conductivity of the nanosheet. Transmission electron microscopy (TEM, FEI Talos F200, Hillsboro, OR, USA) was used to characterize the device's component dimensions and crystal structure. Energy-dispersive spectroscopy (EDS) was used to determine the distribution of various elements in the device. Nano-beam diffraction (NBD) was used to analyze the crystal structure of the channel

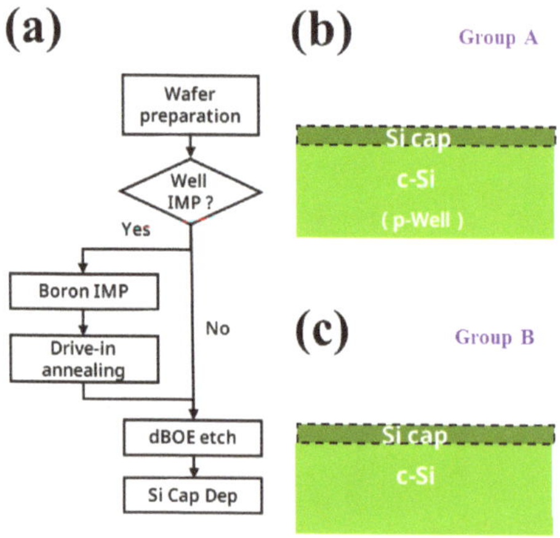

Figure 2. (**a**) Flow diagram of the laser annealing experiment; schematic diagram of (**b**) group A wafer with p-well after Si cap growth; (**c**) group B wafer without p-well after Si cap growth.

3. Results and Discussion

3.1. Structural Analysis of the Si Cap Film Based on RTCVD

Figure 3a is a SEM image of the sample surface of wafer(B-1) after the "dBOE cleaning 0 s + Si cap growth" step, and it can be seen that the wafer has a very smooth surface. In Figure 3b, there is a layer of a-Si film with a thickness of 38.9 nm on the surface of wafer(B-1), and the contrast between the a-Si film and the single-crystal Si of the substrate is different, proving that the a-Si/c-Si interface exists. This result may be due to the natural oxide layer on the wafer having a blocking effect on the Si (100) crystal plane, suppressing the regular arrangement of Si atoms during the growth of the Si cap.

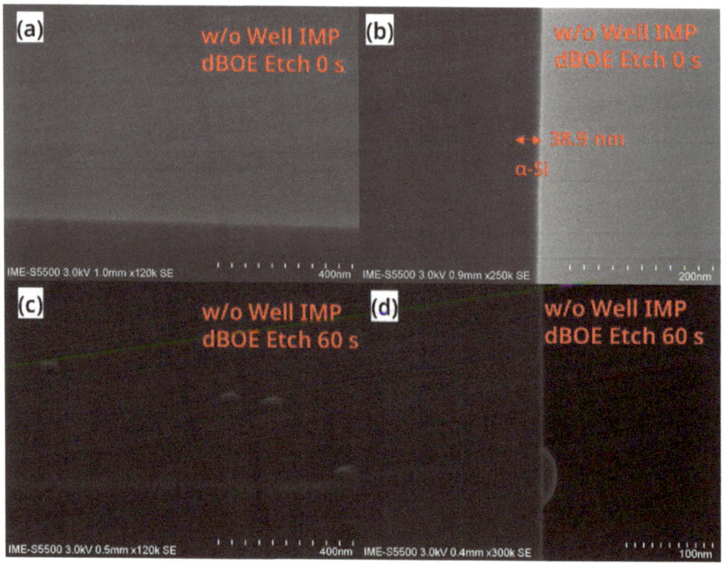

Figure 3. SEM images of (**a**) the surface and (**b**) the cross-sectional view of the blank wafer after the "dBOE cleaning 0 s + Si cap growth" process; SEM images of (**c**) the surface and a (**d**) cross-sectional view of the wafer after the "dBOE cleaning 60 s + Si cap growth" process.

Figure 3c is a SEM image of the wafer(B-2) surface after the "dBOE cleaning 60 s + Si cap growth" step, and Figure 3d is its cross-sectional SEM image. In Figure 3d, it can be seen that the interface of a-Si/c-Si is not visible after 60 s of BOE cleaning. This result indicates that the Si (100) crystal surface could act as a seed layer. Meanwhile, as the RTCVD chamber was designed for the deposition of a-Si and poly-Si thin films, a small number of particles in the equipment may cause some hillock-like defects such as those in Figure 3c during the growth of the Si cap.

In addition, AFM tests were carried out on the wafer(B-1) and wafer(B-2) surfaces. In Figure 4a,b, the root mean square roughness (RMS) of the wafer(B-1) surface is 0.50 nm, and the RMS of the wafer(B-2) surface is 3.65 nm. This result is due to some small bulges on the surface in Figure 4b, increasing the RMS of the entire region. The RMS of the non-bulge area on the wafer(B-2) surface is relatively low. These test results show that the pre-clean step significantly impacts the morphology of the Si cap.

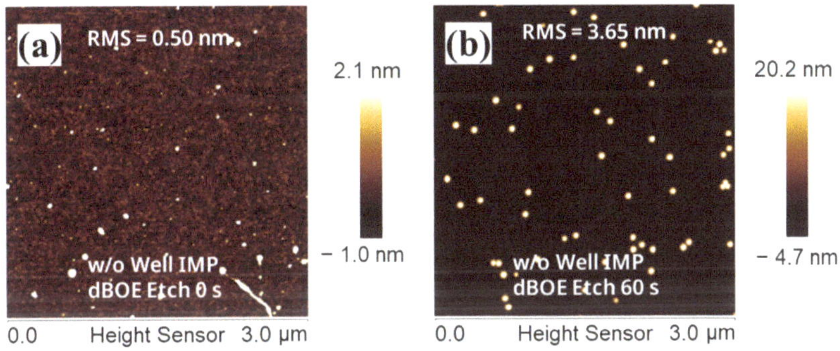

Figure 4. AFM images of the surface of (**a**) the blank wafer after the "dBOE cleaning 0 s + Si cap growth" process and (**b**) the blank wafer after the "dBOE cleaning 60 s + Si cap growth" process.

3.2. Effect of the Laser Annealing Process on the Nanosheet with the RC Channel

As shown in Figure 1g, we performed a laser annealing experiment on the vertical nanosheet with an "exposed top" structure. Figure 5a is the SEM image of the nanosheet surface before laser annealing, and Figure 5b is the SEM image of the vertical nanosheet surface after laser annealing. The energy density of the laser used was 1.67 J/cm^2. The above results show that the top silicon of the "exposed vertical nanosheet" changed from having a right-angled surface to a curved surface. This result shows that the top silicon of the nanosheet underwent a recrystallization process of "a-Si (solid)-Si (liquid)-c-Si (solid)" [20,21]. Moreover, it can be seen that the surface morphology of the ring-shaped recrystallized nanosheet was relatively uniform, and no apparent cracks appeared.

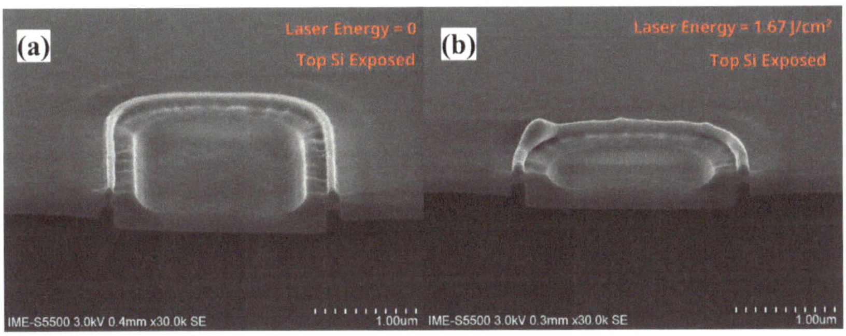

Figure 5. SEM image of the surface of the "exposed vertical nanosheet" (**a**) before and (**b**) after the laser annealing process.

Next, we also studied the effect of different laser energy densities on the recrystallization process of the vertical nanosheets. Figure 6a is a top-view SEM of the vertical nanosheet before laser annealing. The dishing pits in the center of the circular nanosheet caused by the CMP process can be observed. We used A, B, C, and D, four groups of lasers with different energy densities, to irradiate the "exposed vertical nanosheet" (the laser energy is 0, 1.33, 1.67 and 2.00 J/cm^2). Obtained after the laser annealing process, the AFM test results of the above four groups of samples are shown in Figure 6b–e. It can be seen from Figure 6b–e that when the laser energy density was 1.67 J/cm^2, the color difference between the nanosheet top silicon and its surrounding HARP oxide was the smallest, which means that the height difference was the smallest. This indicates that the top silicon of the nanosheet shrunk significantly due to the recrystallization process under this annealing condition. At the same time, when the laser energy density was 2.00 J/cm^2, the color difference between the top silicon of the nanosheet and the surrounding HARP oxide began to increase, which means that the roughness of the HARP oxide began to increase significantly. This result indicates that the energy of the "2.00 J/cm^2" laser was too high and began to have an ablation effect on the wafer surface.

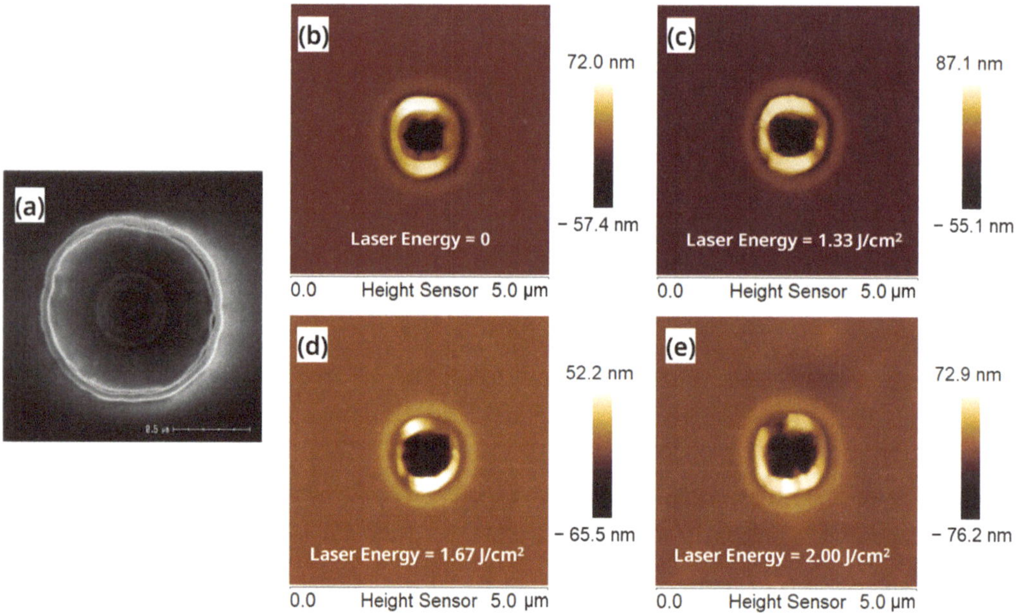

Figure 6. (**a**) Top-view SEM of the surface of the "exposed vertical nanosheet" before the laser annealing process; AFM images of the surface of the "exposed vertical nanosheet" under the laser energy of (**b**) 0 J/cm^2, (**c**) 1.33 J/cm^2, (**d**) 1.67 J/cm^2 and (**e**) 2.00 J/cm^2.

In addition, we also carried out a C-AFM test on the four groups of samples, A, B, C and D. With the increase in the laser energy density, the tunneling current of nanosheets first increased and then decreased, as shown in Figure 7a–d. This indicates that the laser with the energy density of 1.67 J/cm^2 was the most favorable for the recrystallization. Under the condition of 1.67 J/cm^2, the energy absorbed by a-Si from the laser was enough to melt itself, and the ablation effect caused by high laser energy was avoided.

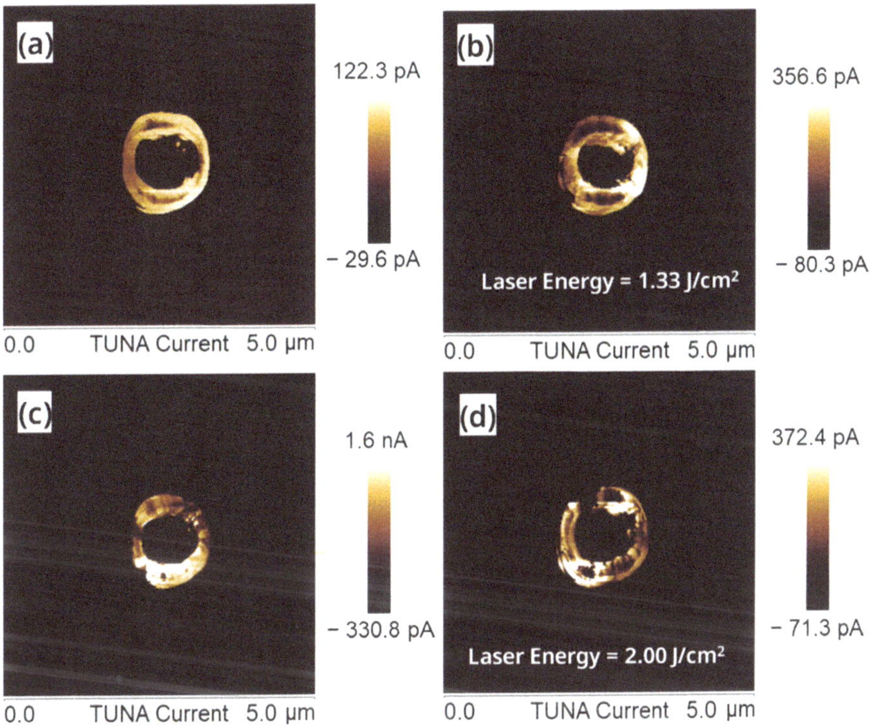

Figure 7. C-AFM images of the surface of the "exposed vertical nanosheet" under the laser energy of (**a**) 0 J/cm^2, (**b**) 1.33 J/cm^2, (**c**) 1.67 J/cm^2 and (**d**) 2.00 J/cm^2.

3.3. Crystal Structure Analysis of the Vertical Nanosheet before Laser Annealing

Next, we analyzed the changes in the crystal structure of the vertical nanosheet before and after laser annealing by means of TEM. Figure 8a is the TEM test result of the cross-section of the nanosheet before laser annealing. The C-shaped Si cap at the bottom of the channel and the single-crystal Si substrate have a black contrast, which indicates that the Si cap grown on Si (100) by RTCVD was a single-crystal structure. Simultaneously, the top silicon of the vertical nanosheet and the upper half of the C-shaped channel had a light contrast, indicating that the silicon in these regions was amorphous. Using Figure 8a, the Si cap thickness can be measured. The thickness of the Si cap grown on c-Si, SiN and a-Si was 10.4 nm, 13.1 nm and 23.1 nm, respectively. This is a clear deviation from the expected growth thickness. The different growth rates of Si caps on these interfaces were due to the differences in their respective surface chemical reaction rates. Figure 8b–d shows the HRTEM images of the three regions of the nanosheet in Figure 8a. In Figure 8b, the lattice diffraction signal cannot be observed in the upper half of the C-shaped channel, and the FFT image in Figure 8e shows a dispersed circle. These results indicate that the Si cap in this region was amorphous, like the HARP oxide in Figure 8b. The regions in Figure 8c,d are all in black contrast, and both have Si (111) plane-aligned twin dislocations. At the same time, the FFT results in Figure 8f,g also show a diamond-shaped pattern of the Si (110) crystal plane. These results show that the Si cap grown near the Si (100) seed layer had a single-crystal structure, but the annealing process is required to eliminate defects such as twin dislocations.

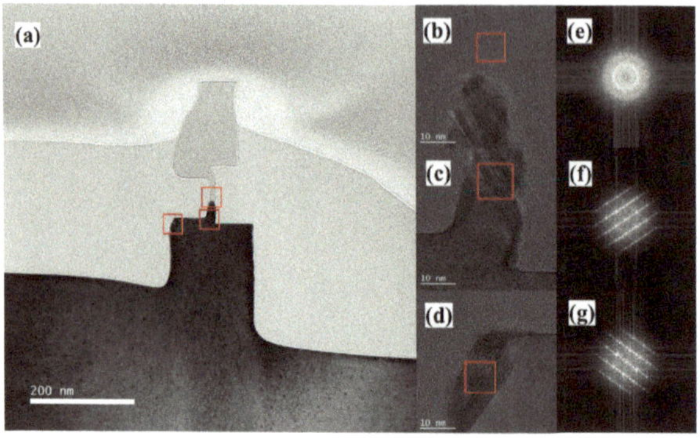

Figure 8. (**a**) TEM image (the red boxs refer to the area of HRTEM images), (**b**–**d**) HRTEM images (the red boxs refer to the area for FFT images), and (**e**–**g**) FFT images of the cross-section of the "exposed vertical nanosheet" before the laser annealing process.

Figure 9a–c is the HAADF-STEM image of the cross-section of the vertical nanosheet before laser annealing, and the EDS mapping images of O and Si elements, respectively. Figure 9d,e are the EDS line scan results of the red dotted line in Figure 9a. In the HAADF-STEM image, there is an obvious interface between the Si cap layer and the bottom c-Si seed layer before laser annealing, and the twin dislocations in the lower left corner of the C-shaped channel are in bright-white contrast. This shows that although the Si cap film grown by RTCVD could form a structure close to that of a single crystal with the assistance of the c-Si seed layer, the film may still have had some lattice defects. These defects may need to be repaired via a laser annealing process. The EDS line scan results in Figure 9d show that there was no significant oxygen element at the Si cap/c-Si interface, which indicates that the "dBOE 60 s cleaning" process removed the natural oxide layer on the surface of the c-Si seed layer, enabling the growth of the single-crystal Si cap on the seed. According to the curve of the Si element in Figure 9e, the thickness of the Si cap was about 13.4 nm, which is basically consistent with the results in Figure 8a.

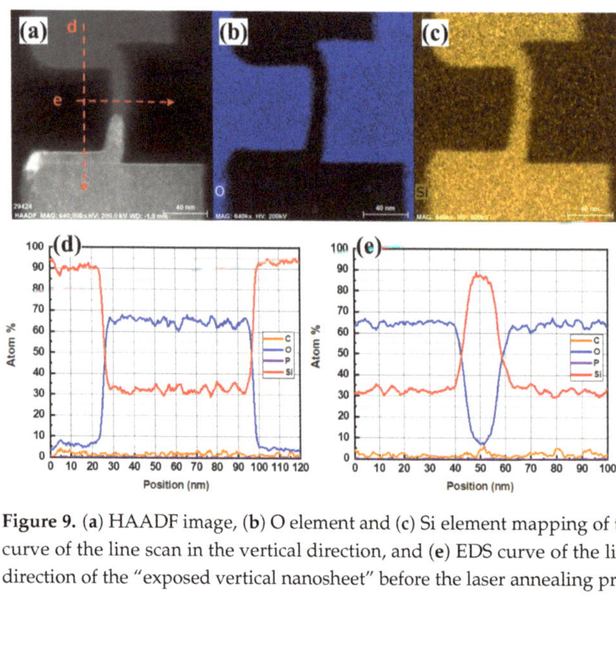

Figure 9. (**a**) HAADF image, (**b**) O element and (**c**) Si element mapping of the cross-section, (**d**) EDS curve of the line scan in the vertical direction, and (**e**) EDS curve of the line scan in the horizontal direction of the "exposed vertical nanosheet" before the laser annealing process.

3.4. Crystal Structure Analysis of the Vertical Nanosheet after Laser Annealing

Figure 10a is the TEM image of the vertical nanosheet after laser annealing. Compared with to sample before laser annealing, the TEM images of the C-shaped channel and the top silicon show the black contrast of the single crystal. In Figure 10b,e, twin dislocations exist in the middle region of the sample channel after annealing. In Figure 10c,d, compared to the samples before annealing, the twin dislocations in these regions of the samples disappeared after annealing, indicating that the laser annealing process can repair these dislocation defects. Next, as shown in Figure 11a–d, the three regions of the nanosheet channel were tested via nanobeam diffraction (NBD), and the spot size of the electron beam was 0.45 nm. The results show that the three regions of the channel all exhibit diffraction patterns of the Si (110) plane index.

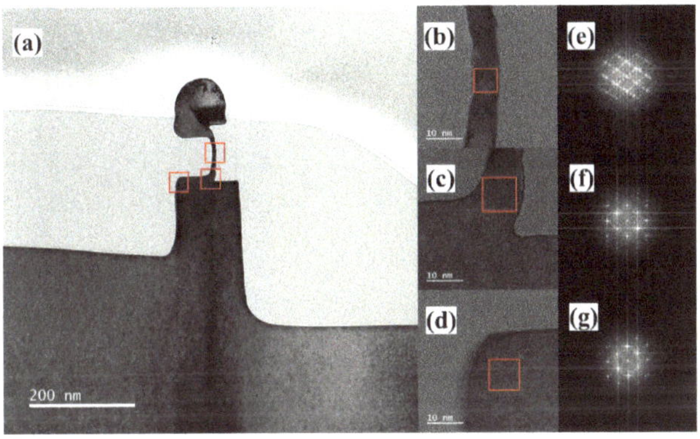

Figure 10. (**a**) TEM image (the red boxs refer to the area of HRTEM images), (**b**–**d**) HRTEM images (the red boxs refer to the area for FFT images), and (**e**–**g**) FFT images of the cross-section of the "exposed vertical nanosheet" after the laser annealing process.

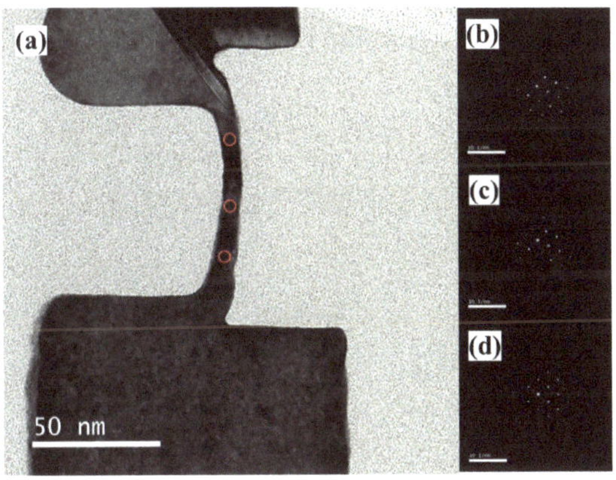

Figure 11. (**a**) TEM image (the red circles refer to the area of NBD images) and (**b**–**d**) NBD images of the cross-section of the "exposed vertical nanosheet" after the laser annealing process.

Figure 12a–c is the HAADF-STEM image of the cross-section of the vertical nanosheet after the laser annealing process and the EDS mapping images of the O element and the Si element. Figure 12d,e shows the EDS line scan results of the red dotted line in Figure 12a. From the HADDF diagram, the Si cap/c-Si interface of the sample after annealing can be seen. At the same time, it can be seen that the bright spot of the twin dislocations in the lower left corner of the nanosheet disappeared. This shows that the laser annealing process has a good repair effect on lattice defects. In Figure 12b,d, the interface of Si cap/c-Si has no oxygen enrichment. Via the comparison of Figures 12e and 9e, it can be seen that the thickness of the middle channel of the vertical nanosheet with the "exposed top" structure was reduced from 13.4 to 7.0 nm after the laser annealing process. This may be due to the compressive stress exerted by the HARP oxide on the circular vertical nanosheet. During the laser annealing process, the a-Si changed from being in a solid state to a free-flowing liquid Si when the nanosheet absorbed laser energy. At this time, the liquid Si moved upward, the compressive stress on the vertical nanosheet channel began to release, and the surrounding HARP oxide was displaced, which finally reduced the thickness of the nanosheet.

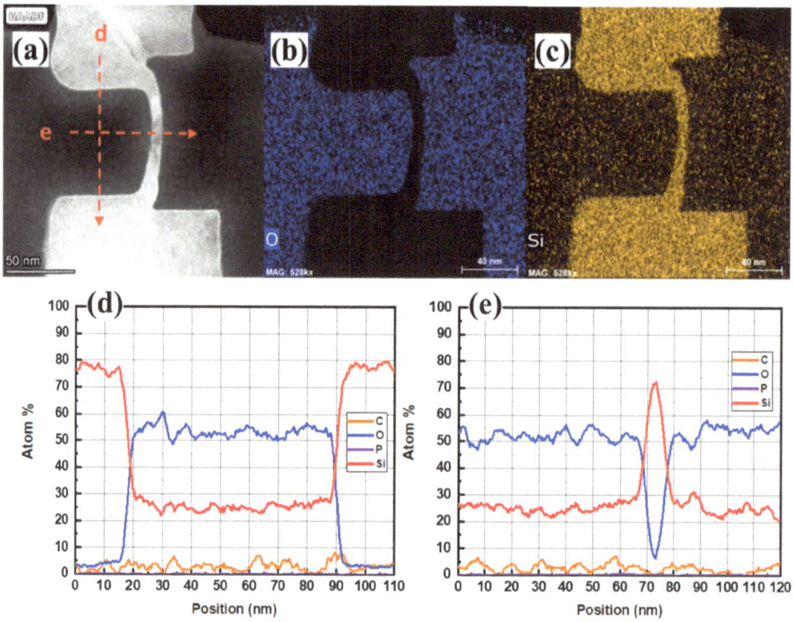

Figure 12. (**a**) HAADF image, (**b**) O element and (**c**) Si element mapping of the cross-section, (**d**) EDS test curve of line scan in vertical direction, and (**e**) EDS test curve of line scan in horizontal direction of the "exposed vertical nanosheet" before the laser annealing process.

In order to further characterize the recrystallization effect of the laser annealing process on larger-sized devices, we also conducted laser annealing experiments on the "exposed vertical nanosheet" with a size of 4 μm × 4 μm and performed a top-view-TEM test analysis. Figure 13a is a schematic diagram of the structure of the TEM sample preparation area. The red-framed part is the FIB slice sample, and the sample thickness was about 100 nm.

Figure 13b is the TEM test result of the sample's top view. The channel with a size of 4 μm × 4 μm remained intact and continuous after laser annealing, which proves the uniformity of the recrystallization process of the "exposed vertical nanosheet". In the HRTEM and FFT images of Figure 14A–D, the channels at the four corners of the sample are all single-crystal structures. In Figure 13e–h, the HAADF images also prove the integrity and continuity of the channel with a size of 4 μm × 4 μm. In Figure 14E–H, the average projected width of the nanosheet channel is about 13.5 nm. This thickness indicates that the RC-VCNFETs device fabricated by the laser annealing process has superior gate control capability.

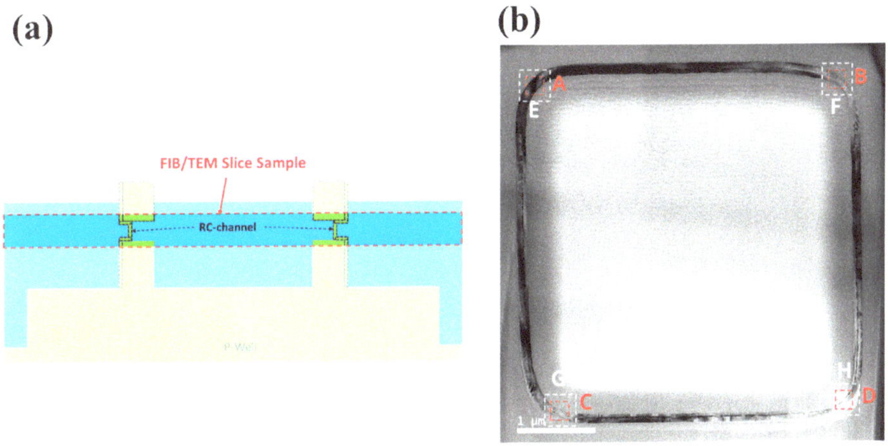

Figure 13. (**a**) Schematic diagram of the FIB sample and (**b**) top-view TEM image of the "exposed vertical nanosheet" after the laser annealing process (the red boxes A–D refer to the area of HRTEM images, and the white boxes E–H refer to the area of HADDF images).

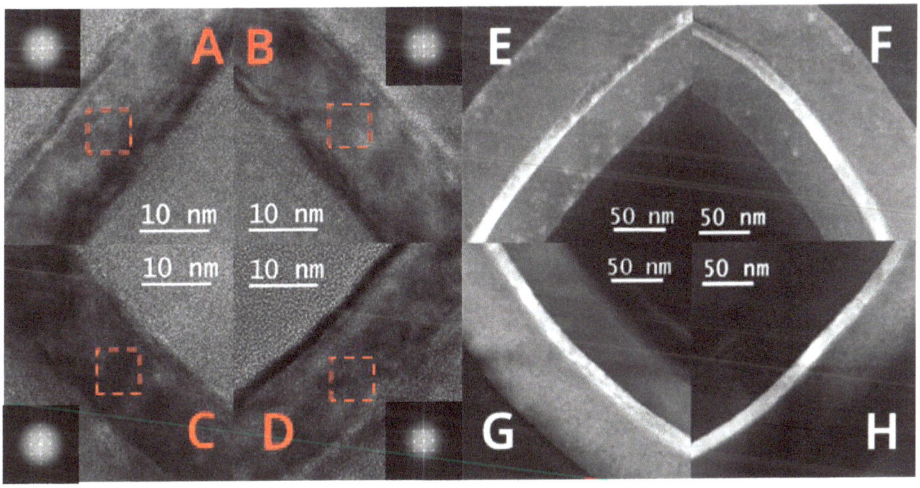

Figure 14. Top view of (**A–D**) HRTEM (the red boxs refer to the area for FFT images) and (**E–H**) HAADF images of the "exposed vertical nanosheet" after the laser annealing process.

3.5. Electrical Properties of the "Exposed-Top" RC-VCNFET Device

Finally, the RC-VCNFET device was fabricated successfully. Figures 15a and 15b,c are the TEM images of the device's top view and cross-section view, respectively. It can be seen from Figure 15a that the upper half of the ring-shaped RC-VCNFET device is a double-gate device, while the lower half of the ring-shaped device is a single-gate device. Next, the electrical test was performed on the "exposed-top" RC-VCNFETs device, and the results are shown in Figure 15d. The I_{on} of this RC-VCNFET device is 11.5 μA/μm (I_D @ $V_{OV} = V_G - V_T = 1$ V, $V_{DS} = 0.65$ V). The SS of the device is 67.0 mV/dec, and the DIBL of the device is 46.7 mV/V. These test results indicate that the performance of the RC-VCNFET needs to be further optimized.

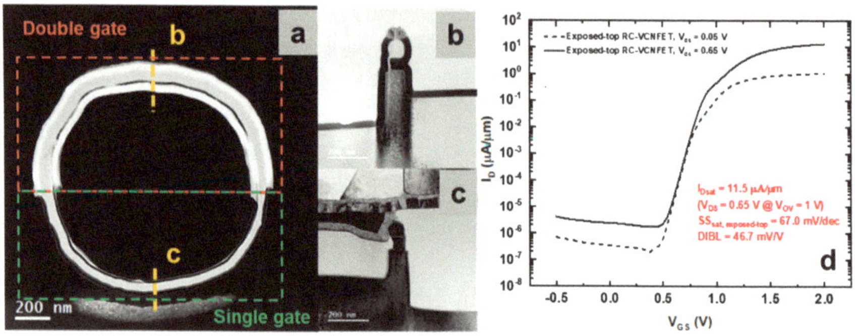

Figure 15. (**a**) The Top View TEM image in dark field, (**b**,**c**) the Cross-section TEM in bright field of the "exposed-top" RC-VCNFET device, and (**d**) the I_D-V_{GS} curves of the "exposed-top" RC-VCNFET.

4. Conclusions

This paper introduces the effect of pre-cleaning conditions on the surface of the wafer on the Si cap film grown by RTCVD. SEM and AFM results revealed that an amorphous Si cap grew on the single-crystal silicon without the pre-clean step. In addition, the laser annealing process was carried out on the vertical nanosheet with an "exposed top" structure, and the crystal structure of the vertical nanosheet before and after laser annealing was characterized by means of SEM, AFM, C-AFM, TEM and NBD. Finally, the high-quality recrystallized vertical nanosheet structure was successfully fabricated, which laid a certain foundation for the preparation of high-performance and low-cost vertical-channel devices in the future.

Author Contributions: Conceptualization, Z.C. and H.Z.; data curation, Z.C.; funding acquisition, H.Z.; project administration, H.Z.; supervision, H.Z. and G.W.; writing—original draft, Z.C.; writing—review and editing, Z.C., H.Z. and G.W.; device process integration, Z.C., Q.W., Z.X., Y.Z. (Yongkui Zhang), S.L., Y.D., Z.Y. and Y.Z. (Yantong Zheng); laser annealing process tuning, J.L.; RTCVD process tuning, W.X.; pre-clean process and ALD process tuning, Z.K.; TEM measurement, A.D.; AFM and C-AFM measurement, J.Y. All authors have read and agreed to the published version of the manuscript.

Funding: This work was supported by the National Natural Science Foundation of China (grant no. 92064002), the Beijing Superstring Academy of Memory Technology (no. E2DF06x003).

Data Availability Statement: The data presented in this study are available upon request from the corresponding author.

Conflicts of Interest: The authors declare no conflict of interest.

References

1. Zhang, S. Review of Modern Field Effect Transistor Technologies for Scaling. *J. Phys. Conf. Ser.* **2020**, *1617*, 012054. [CrossRef]
2. Bae, G.; Bae, D.-I.; Kang, M.; Hwang, S.M.; Kim, S.S.; Seo, B.; Kwon, T.Y.; Lee, T.J.; Moon, C.; Choi, Y.M.; et al. 3 nm GAA Technology Featuring Multi-Bridge-Channel FET for Low Power and High Performance Applications. In Proceedings of the 2018 IEEE International Electron Devices Meeting (IEDM), San Francisco, CA, USA, 1–5 December 2018; pp. 28.7.1–28.7.4.
3. Song, T.; Kim, H.; Rim, W.; Jung, H.; Park, C.; Lee, I.; Baek, S.; Jung, J. A 3-Nm Gate-All-Around SRAM Featuring an Adaptive Dual-Bitline and an Adaptive Cell-Power Assist Circuit. *IEEE J. Solid-State Circuits* **2022**, *57*, 236–244. [CrossRef]
4. Huang, C.-Y.; Dewey, G.; Mannebach, E.; Phan, A.; Morrow, P.; Rachmady, W.; Tung, I.-C.; Thomas, N.; Alaan, U.; Paul, R.; et al. 3-D Self-Aligned Stacked NMOS-on-PMOS Nanoribbon Transistors for Continued Moore's Law Scaling. In Proceedings of the 2020 IEEE International Electron Devices Meeting (IEDM), San Francisco, CA, USA, 12–18 December 2020; pp. 20.6.1–20.6.4.
5. Liu, M. 1.1 Unleashing the Future of Innovation. In Proceedings of the 2021 IEEE International Solid-State Circuits Conference (ISSCC), San Francisco, CA, USA, 13–22 February 2021; Volume 64, pp. 9–16.
6. Park, S.-K. Technology Scaling Challenge and Future Prospects of DRAM and NAND Flash Memory. In Proceedings of the 2015 IEEE International Memory Workshop (IMW), Monterey, CA, USA, 17–20 May 2015; pp. 1–4.

7. Yang, C.-M.; Wei, C.-K.; Chang, Y.J.; Wu, T.-C.; Chen, H.-P.; Lai, C.-S. Suppression of Row Hammer Effect by Doping Profile Modification in Saddle-Fin Array Devices for Sub-30-Nm DRAM Technology. *IEEE Trans. Device Mater. Reliab.* **2016**, *16*, 685–687. [CrossRef]
8. Lee, S.-W.; Kim, S.-Y.; Hwang, K.-M.; Jin, I.K.; Hur, J.; Kim, D.-H.; Son, J.W.; Kim, W.-K.; Choi, Y.-K. A Comprehensive Study of a Single-Transistor Latch in Vertical Pillar-Type FETs with Asymmetric Source and Drain. *IEEE Trans. Electron Devices* **2018**, *65*, 5208–5212. [CrossRef]
9. Cho, Y.S.; Choi, P.H.; Kim, K.H.; Park, J.M.; Hwang, Y.S.; Hong, H.S.; Lee, K.P.; Choi, B.D. Stretched Tunnelling Body Contact Structure for Suppressing the FBE in a Vertical Cell DRAM. *Electron. Lett.* **2019**, *55*, 1252–1253. [CrossRef]
10. Cho, Y.; Kim, H.; Jung, K.; Kim, B.; Hwang, Y.; Hong, H.; Choi, B. Suppression of the Floating-Body Effect of Vertical-Cell DRAM with the Buried Body Engineering Method. *IEEE Trans. Electron Devices* **2018**, *65*, 3237–3242. [CrossRef]
11. Cho, Y.; Choi, P.; Hyeon, Y.; Song, J.; Hwang, Y.; Choi, B. Novel Band-to-Band Tunneling Body Contact (BTBC) Structure to Suppress the Floating-Body Effect in a Vertical-Cell DRAM. *IEEE Electron Device Lett.* **2018**, *39*, 1860–1863. [CrossRef]
12. Jagannathan, H.; Anderson, B.; Sohn, C.-W.; Tsutsui, G.; Strane, J.; Xie, R.; Fan, S.; Kim, K.I.; Song, S.; Sieg, S. Vertical-Transport Nanosheet Technology for CMOS Scaling beyond Lateral-Transport Devices. In Proceedings of the 2021 IEEE International Electron Devices Meeting (IEDM), San Francisco, CA, USA, 11–16 December 2021; p. 26-1.
13. Tsutsui, G.; Song, S.; Strane, J.; Xie, R.; Qin, L.; Zhang, C.; Schmidt, D.; Fan, S.; Hong, B.; Jung, Y. Hardware Based Performance Assessment of Vertical-Transport Nanosheet Technology. In Proceedings of the IEEE International Electron Devices Meeting, San Francisco, CA, USA, 3–7 December 2022.
14. Chang, C.-H.; Chang, V.S.; Pan, K.H.; Lai, K.T.; Lu, J.H.; Ng, J.A.; Chen, C.Y.; Wu, B.F.; Lin, C.J.; Liang, C.S. Critical Process Features Enabling Aggressive Contacted Gate Pitch Scaling for 3nm CMOS Technology and Beyond. In Proceedings of the 2022 International Electron Devices Meeting (IEDM), San Francisco, CA, USA, 3–7 December 2022; pp. 27.1.1–27.1.4.
15. Yin, X.; Zhang, Y.; Zhu, H.; Wang, G.L.; Li, J.J.; Du, A.Y.; Li, C.; Zhao, L.H.; Huang, W.X.; Yang, H. Vertical Sandwich Gate-All-around Field-Effect Transistors with Self-Aligned High-k Metal Gates and Small Effective-Gate-Length Variation. *IEEE Electron Device Lett.* **2019**, *41*, 8–11. [CrossRef]
16. Zhang, Y.; Ai, X.; Yin, X.; Zhu, H.; Yang, H.; Wang, G.L.; Li, J.J.; Du, A.Y.; Li, C.; Huang, W.X. Vertical Sandwich GAA FETs with Self-Aligned High-k Metal Gate Made by Quasi Atomic Layer Etching Process. *IEEE Trans. Electron Devices* **2021**, *68*, 2604–2610. [CrossRef]
17. Li, C.; Zhu, H.; Zhang, Y.; Wang, Q.; Yin, X.; Li, J.; Wang, G.; Kong, Z.; Ai, X.; Xie, L. First Demonstration of Novel Vertical Gate-All-around Field-Effect-Transistors Featured by Self-Aligned and Replaced High-κ Metal Gates. *Nano Lett.* **2021**, *21*, 4730–4737. [CrossRef] [PubMed]
18. Xiao, Z.R.; Wang, Q.; Zhu, H.L.; Chen, Z.; Zhang, Y.K.; Li, J.J.; Zhou, N.; Gao, J.F.; Ai, X.Z.; Lu, S.S.; et al. Vertical C-Shaped-Channel Nanosheet FETs Featured with Precise Control of Both Channel-Thickness and Gate-Length. *IEEE Electron Device Lett.* **2022**, *43*, 1183–1186. [CrossRef]
19. Xiao, Z.R.; Zhu, H.L.; Wang, Q.; Chen, Z.; Liu, Z.Y.; Zhang, Y.K.; Yan, Z.J.; Shi, Y.F.; Zhou, N.; Li, J.J.; et al. Vertical N-Type and P-Type Nanosheet FETs with C-Shaped Channel. *IEEE Trans. Electron Devices* **2023**, *70*, 1380–1385. [CrossRef]
20. Liu, Y.-W.; Hu, H.-W.; Hsieh, P.-Y.; Chung, H.-T.; Chang, S.-J.; Liu, J.-H.; Huang, P.-T.; Yang, C.-C.; Shen, C.-H.; Shieh, J.-M. Single-Crystal Islands (SCI) for Monolithic 3-D and Back-End-of-Line FinFET Circuits. *IEEE Trans. Electron Devices* **2021**, *68*, 5257–5262. [CrossRef]
21. Son, Y.-I.; Shin, J. Numerical Study on the Laser Annealing of Silicon Used in Advanced V-NAND Device. *Materials* **2022**, *15*, 4201. [CrossRef] [PubMed]

nanomaterials

MDPI

Article

Investigation on Recrystallization Channel for Vertical C-Shaped-Channel Nanosheet FETs by Laser Annealing

Zhuo Chen [1,2], Huilong Zhu [1,*], Guilei Wang [3], Qi Wang [1], Zhongrui Xiao [1,2], Yongkui Zhang [1], Jinbiao Liu [1,2], Shunshun Lu [1], Yong Du [1], Jiahan Yu [1,2], Wenjuan Xiong [1], Zhenzhen Kong [1,2], Anyan Du [1], Zijin Yan [1,2] and Yantong Zheng [1,2]

1 Key Laboratory of Microelectronics Devices & Integrated Technology, Institute of Microelectronics, Chinese Academy of Sciences, Beijing 100029, China; chenzhuo@ime.ac.cn (Z.C.); wangqi@ime.ac.cn (Q.W.); xiaozhongrui@ime.ac.cn (Z.X.); zhangyongkui@ime.ac.cn (Y.Z.); liujinbiao@ime.ac.cn (J.L.); lushunshun@ime.ac.cn (S.L.); duyong@ime.ac.cn (Y.D.); yujiahan@ime.ac.cn (J.Y.); xiongwenjuan@ime.ac.cn (W.X.); kongzhenzhen@ime.ac.cn (Z.K.); duanyan@ime.ac.cn (A.D.); yanzijin@ime.ac.cn (Z.Y.); zhengyantong@ime.ac.cn (Y.Z.)

2 Microelectronics Institute, University of Chinese Academy of Sciences, Beijing 100049, China

3 Process Integration, Beijing Superstring Academy of Memory Technology, Beijing 100176, China; guilei.wang@bjsamt.org.cn

* Correspondence: zhuhuilong@ime.ac.cn

Abstract: Transistor scaling has become increasingly difficult in the dynamic random access memory (DRAM). However, vertical devices will be good candidates for $4F^2$ DRAM cell transistors (F = pitch/2). Most vertical devices are facing some technical challenges. For example, the gate length cannot be precisely controlled, and the gate and the source/drain of the device cannot be aligned. Recrystallization-based vertical C-shaped-channel nanosheet field-effect transistors (RC-VCNFETs) were fabricated. The critical process modules of the RC-VCNFETs were developed as well. The RC-VCNFET with a self-aligned gate structure has excellent device performance, and its subthreshold swing (SS) is 62.91 mV/dec. Drain-induced barrier lowering (DIBL) is 6.16 mV/V.

Keywords: vertical channel transistor; self-aligned; laser annealing; recrystallization; Si cap

1. Introduction

Recently, the technology for the mass production of logic devices has evolved to the 3 nm technology node [1]. In the future, Intel, Samsung, and TSMC will continue to optimize the power, performance, area, and cost (PPAC) for the logic device by using new technologies at the 2 nm technology node, such as gate-all-around FETs (GAAFETs) [2,3] and buried power rail (BPR) [4–8]. However, the scaling of the lateral device has become more and more difficult, and the cost of tape-out has become unaffordable for major design houses. At the same time, vertical devices will be competitive candidates for $4F^2$ cell transistors in the future DRAM device [9–13]. There are many research reports on vertical devices, which can be divided into two routes. The "bottom-up" route enables the growth of vertical nanowire channels by using metal nanoparticle-induced catalysis [14,15]. However, there is a problem with metal elements, such as Au contamination, so it is not compatible with the standard CMOS process. In addition, the "top-down" approach to fabricating vertical transistor devices through lithography and etching processes has been reported by Samsung and IBM [16,17]. However, there are some problems with this route. For example, the device gate length and channel thickness are challenging to control precisely, and the gate cannot be aligned with the source/drain of the vertical device in this route.

In order to solve the above problems, vertical sandwich gate-all-around (GAA) FETs (VSAFETs) based on the SiGe channel are proposed, which have a self-aligned structure between the gate and the source/drain [18–21]. Recently, vertical C-shaped-channel nanosheet field-effect transistors (VCNFETs) based on the epitaxial process were reported [22,23]. In the process flow of the VCNFETs, the C-shaped SiGe cavity, which is used as a seed layer for the epitaxial growth of the vertical Si channel, is formed through steps of the Si/SiGe/Si superlattice epitaxy and the SiGe selective etching. In this way, the epitaxial process can precisely control the channel thickness of VCNFETs. The device has excellent gate control and current drive performance among similar vertical devices. However, these devices require the expensive Si and SiGe epitaxial processes in the above works. In particular, there are two epitaxial processes in the process flow of the VCNFETs. At the same time, the device introduces Ge elements in the front-end-of-line (FEOL) process, which will increase additional costs in the contamination control of the process equipment.

Therefore, this work proposes a recrystallization-based vertical C-shaped-channel nanosheet field-effect transistors (RC-VCNFETs) and its process integration method. In this method, the substrate Si of the (100) crystal plane is used as a seed layer, and a Si cap layer is grown by a "rapid thermal chemical vapor deposition (RTCVD)" process. The device is subjected to laser annealing so that the amorphous Si cap layer uses the (100) crystal plane of the substrate as a lattice template and recrystallizes vertically upwards to form a high-quality single-crystal Si channel. This method avoids the problem of expensive epitaxial process and contamination of Ge element. Moreover, the RC-VCNFETs have good electrical properties and perfect gate control capability.

2. Materials and Methods

The RC-VCNFETs were prepared on a p-type Si (100) substrate with a resistivity of 8–12 Ohm·cm, and the main process steps are shown in Figure 1a–h. As shown in Figure 2a, there are mandrels with the following sizes: 1 μm × 1 μm, 2 μm × 2 μm, and 4 μm × 4 μm. Firstly, the p-well of the wafer was formed by high-energy boron ion implantation. An 80 nm silicon nitride (SiN) film was deposited on the crystal silicon (c-Si) wafer surface by plasma-enhanced chemical vapor deposition(AMAT Producer S PECVD, Applied Materials, Santa Clara, CA, USA), followed by 180 nm amorphous silicon (a-Si) deposition at 580 °C, as shown in Figure 3a. Then, 10 nm SiO_2, 300 nm a-Si, and 300 nm SiO_2 films (abbreviated as OSO stacks) were sequentially deposited by- PECVD. The OSO stack was patterned by I-line photolithography and reactive ion etching (RIE) to form the OSO hard mask (HM) shown in Figure 1a. The top 300 nm SiO_2 film in the OSO HM was used as the hard mask, the middle 300 nm a-Si film was used as the mandrel, and the bottom 10 nm SiO_2 was used as the etch stop layer when etching the mandrel by tetramethylammonium hydroxide (TMAH). In the next step, a silicon oxide film was deposited and etched by anisotropy to form silicon oxide sidewalls, as shown in Figure 1b. Then, the OSO HM and oxide spacer were combined into a cap-shaped hard mask.

Furthermore, the a-Si/SiN/c-Si stacks were etched by RIE (TCP9400, Lam Research, Fremont, CA, USA) to form a steep sidewall, as shown in Figure 1c. Afterward, the SiN in the cavity was isotropically etched to form a C-shaped cavity structure, and the c-Si at the bottom of the C-shaped cavity was a seed layer for the next step of the growth of the Si cap by RTCVD (Centura, Applied Materials, Santa Clara, CA, USA). As shown in Figures 1d and 3b, the natural oxide on the surface of the c-Si was removed by the diluted buffered oxide etchant (dBOE). Then, the wafer was immediately loaded into the chamber of the RTCVD equipment, growing a 20 nm Si cap at 580 °C. Meanwhile, the Si cap covering the surface of the OSO cap HM was removed by RIE. Next, the oxide film was deposited with high aspect ratio process. Then, the HARP oxide was polished using chemical mechanical polishing (CMP, FRX200, Ebara, Tokyo, Japan). When HARP was polished to a certain height, the mandrel was exposed. Subsequently, the mandrel was etched by TMAH. Next, 10 nm silicon oxide CESL was etched using RIE. Next, as shown in Figure 1e, the a-Si/SiN/c-Si stacks are etched by RIE. Then, the remaining SiN in the cavity was removed by phosphoric acid solution at 160 °C, as shown in Figure 3c. A 20 nm thick oxide layer was formed on the top of the device with the HARP oxide deposition and the CMP process. As shown in Figure 1f, the wafer surface was irradiated with Nd:YLF pulsed laser (the laser annealing equipment was developed by the Institute of Microelectronics, Chinese Academy of Sciences, the laser's wavelength is 527 nm, the pulse width is 200 ns, and the frequency is 200 Hz, and the energy density is 1.67 J/cm^2). At this time, a-Si turned into liquid instantly and then vertically recrystallized to form single crystal silicon channels, and the top view of the device is shown in Figure 3d. During this process, the c-Si inside the SiN cavity served as a lattice template for a-Si recrystallization. As shown in Figure 1g, through the STI Recess step, the oxide on the surface of the device and active area (AA) were removed. The source and drain electrodes of the device were formed by implanting arsenic at a ten-degree angle and phosphorus at a zero-degree angle. Then the high-k metal gate (HKMG) stacks were grown by the atomic layer deposition (ALD, TFS 200, Espoo, Finland) process. After the i-line lithography and etching process, a part of the metal gate was left on AA to serve as a landing pad for the gate contact hole, whose structure is shown in Figures 2a and 3e. At this point, self-aligned metal gates were formed inside a C-shaped cavity on both sides of the recrystallization channel (RC-channel), as shown in Figures 1h and 3f–i. Then CESL SiN film and TEOS were sequentially deposited. Moreover, the interlayer dielectric (ILD) is formed through the CMP process. Subsequently, through the processes of contact hole etching, Ti/TiN liner, tungsten film deposition, and tungsten CMP, four contact plugs, including contact source (CS), contact drain (CD), contact gate (CG), and contact well (CW) are formed. Finally, as shown in Figure 2b, the entire process flow was completed through the processes of PVD deposition, lithography, etching, and alloy annealing to form the Al Pad for interconnection.

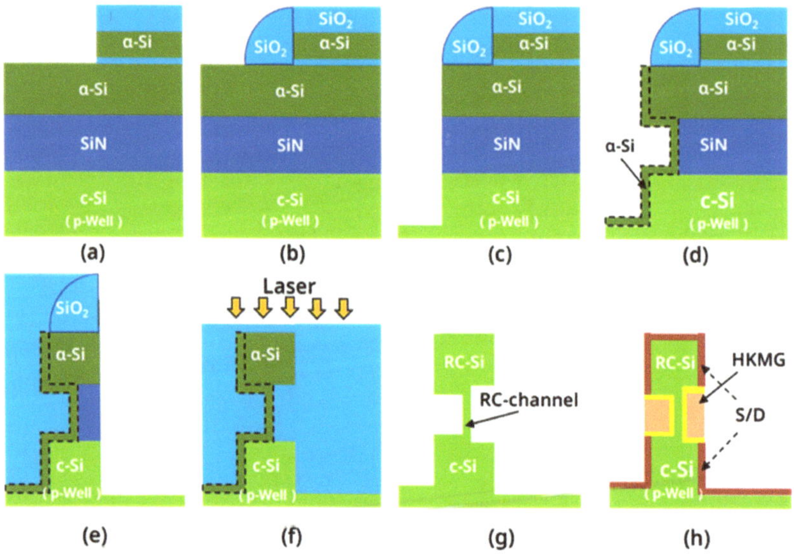

Figure 1. Schematic diagram of the key process steps for the RC-VCNFETs: (**a**) OSO HM formation; (**b**) oxide spacer formation; (**c**) a-Si/SiN/c-Si etching; (**d**) Si cap deposition; (**e**) inner a-Si/SiN/c-Si etching; (**f**) laser annealing; (**g**) STI Recess; (**h**) HKMG etching.

Scanning electron microscopy (SEM) was used to observe the topography of the surface and cross-section of the sample, thereby measuring the film thickness and etching depth. Atomic force microscopy (AFM) was used to evaluate the film surface roughness. Transmission electron microscopy (TEM) characterized the device's component dimensions and crystal structure. Energy-dispersive spectroscopy (EDS) was used to determine the distribution of various elements in the device.

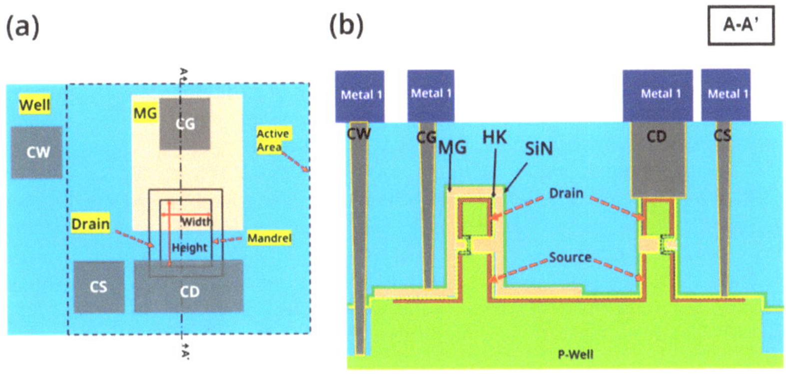

Figure 2. (**a**) Top view and (**b**) cross-sectional view of the device schematic diagram.

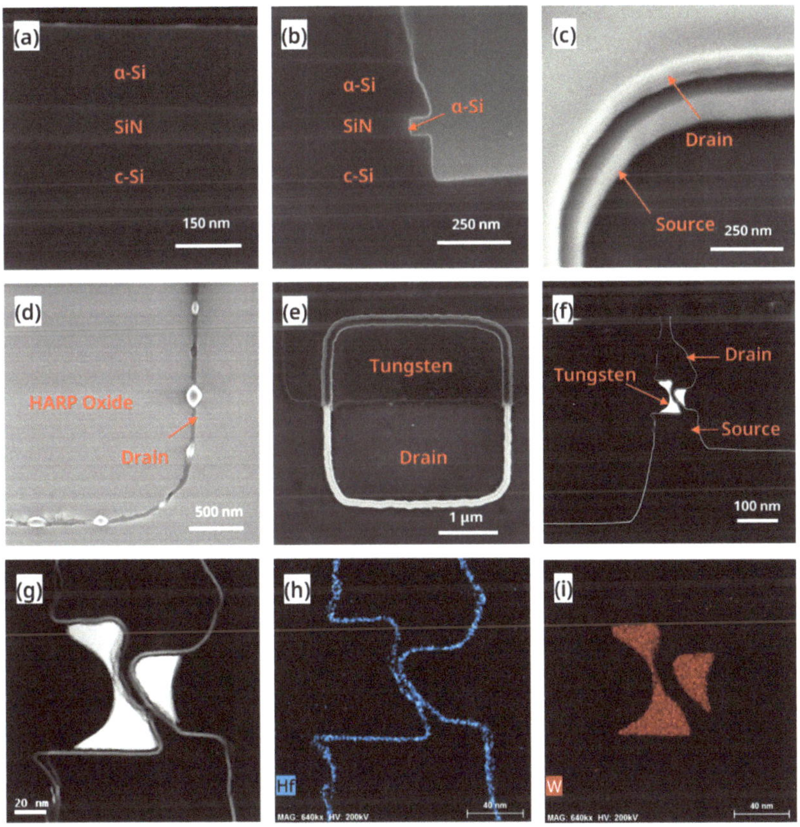

Figure 3. SEM images (**a**) after SiN and a-Si deposition, (**b**) after the growth of Si cap, (**c**) after inner SiN removal, (**d**) after laser annealing (Top view), and (**e**) HKMG etching. (**f**) STEM-HAADF image and (**g**) zoom-in HAADF image of the RC-VCNFET. EDS mapping of (**h**) Hf element and (**i**) W element.

3. Results and Discussion

3.1. Key Process Module—Si Cap Deposition and Pre-Clean

The RTCVD growth process of the Si cap channel is critical as it relates to subsequent recrystallization processes. Based on the experience of Si and SiGe epitaxial processes, pre-treatment processes are essential for the growth of high-quality crystalline Si channels. Before integrating this process module into the entire process flow, we studied the effect of pre-cleaning on the deposition of the Si cap on a Si (100) substrate. Figure 4a,c show the SEM images of the wafers with p-well after dBOE cleaning for 0 s and 60 s, followed by 40 nm Si cap deposition. Under pre-cleaning conditions for 0 s, the sample surface is very smooth. However, the cross-sectional SEM of Figure 4b shows that the substrate surface has a structure with significantly different contrast, which proves the existence of the a-Si/c-Si interface. Meanwhile, under pre-cleaning conditions for the 60 s, some hillock-like structures are on the sample surface. However, no a-Si/c-Si interface exists in Figure 4d. This result indicates that the crystal structure of the deposited Si cap film is so close to the c-Si at the bottom that the secondary electron signal cannot distinguish them.

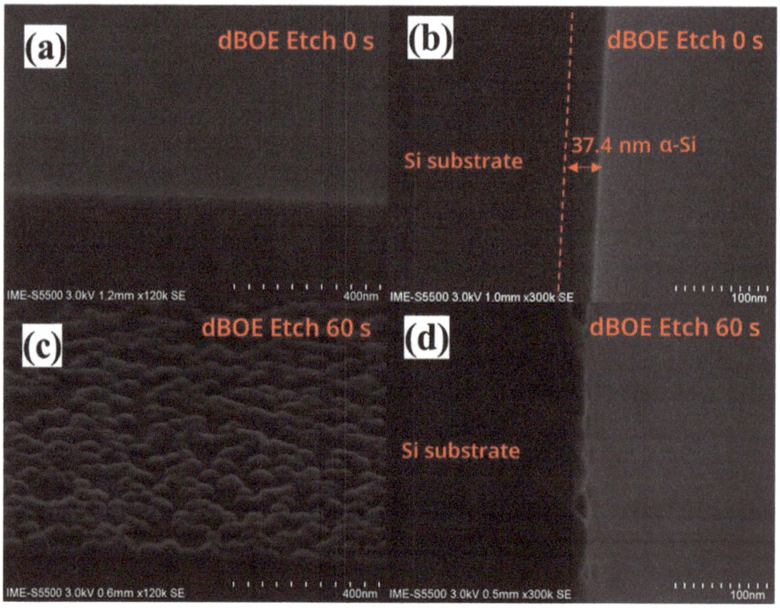

Figure 4. (**a**) Top view SEM image and (**b**) cross-sectional view SEM image of the wafer after the "dBOE cleaning 0 s + Si cap growth" process; (**c**) top view SEM image and (**d**) cross-sectional view SEM image of the wafer after the "dBOE cleaning 60 s + Si cap growth" process.

Figure 5a,b show the AFM images of wafers after 0 s and 60 s of dBOE cleaning, followed by 40 nm Si cap deposition. The RMS of the sample surface pre-cleaned for 60 s is 1.89 nm, and the maximum height of the surface protrusions is 14.9 nm, which shows that although the 60 s dBOE removes the natural oxide layer on the surface of Si (100) wafer. There may still be some defects on the surface of the wafer that cannot be removed by dBOE etching. These defects result in the existence of these hillock-like structures on the surface of the wafer. Some processes, such as well implantation, may introduce these defects. In order to further optimize the quality of the film and the yield of the device in the future, it may be necessary to further clarify the root causes of these defects.

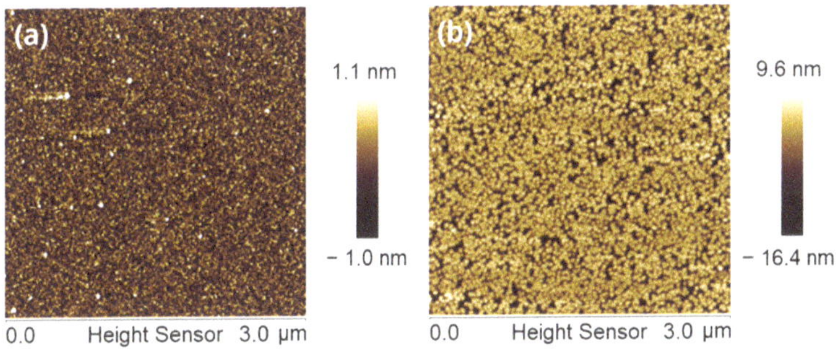

Figure 5. AFM images of the surface of (**a**) the wafer after the "dBOE cleaning 0 s + Si cap growth" process and (**b**) the wafer after the "dBOE cleaning 60 s + Si cap growth" process.

In addition, we also analyzed the structure of the final devices. Figure 6a shows a cross-sectional view TEM image of the RC-VCNFETs, and its structure is shown in Figure 1d. It can be found that the Si cap film deposited by RTCVD has excellent conformality and can well fill the SiN cavity with a width of 80 nm. Figure 6b is a high-resolution transmission electron microscope (HRTEM) image of the sample. A clear Si lattice shows that the Si film that grows on the c-Si seed layer in the SiN cavity is very close to the single crystal structure. As shown in Figure 6c, the Fourier transform is performed on the red box area in Figure 6b, and the diffraction pattern of the silicon (110)-oriented planes is obtained, which is the same as the crystal plane index in the wafer notch direction.

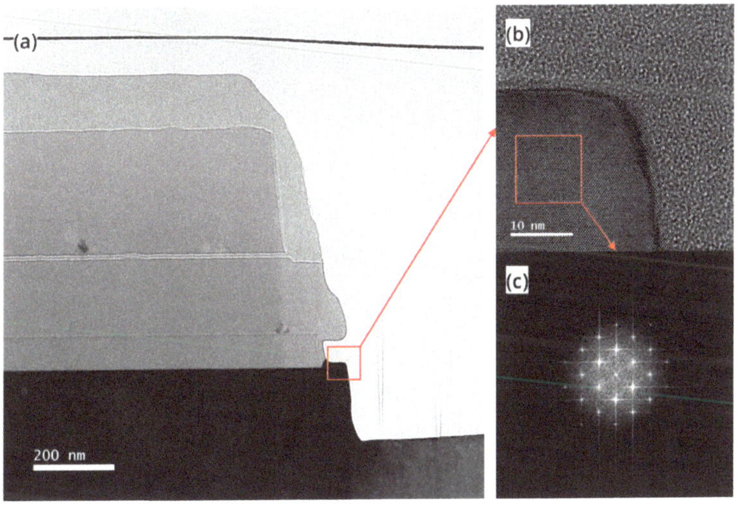

Figure 6. (**a**) TEM image, (**b**) HRTEM image, and (**c**) FFT image of the cross-section of the device after the Si cap growth.

3.2. Structure of RC-Channel

To further evaluate the crystal quality of the RC-channel, we conducted a TEM analysis on the devices after laser annealing. Figure 7a shows that most areas of the RC-VCNFET device contrast similarly with the c-Si substrate in the TEM image, indicating that the RC-channel is high-quality recrystallized Si. HRTEM images in Figure 7b,d also confirm the presence of high-quality recrystallized Si, while Fourier transform results in Figure 7e,g show that this area is a (110) crystal plane. However, as shown in Figure 7a, darker regions at the top of the RC-VCNFET device suggest some defects. As shown in Figure 7c, there is a dislocation along the Si (111) direction at the lower left corner area of the top Si. The Fourier transform result in Figure 7f also indicates twinning defects exist in this area. In addition, it should be emphasized that there are fine cracks in the HARP oxide on the top of the device in Figure 7a, which indicates that the a-Si tends to expand upwards during the process of transitioning from liquid to solid under the pulsed laser, which is consistent with the SEM image in Figure 3d. In Figure 1f, the HARP oxide encapsulating RC-VCNFET device may negatively affect the recrystallization process of the a-Si. Therefore, adjusting HARP oxide height may improve the crystal quality of the Si cap layer in the RC-VCNFETs.

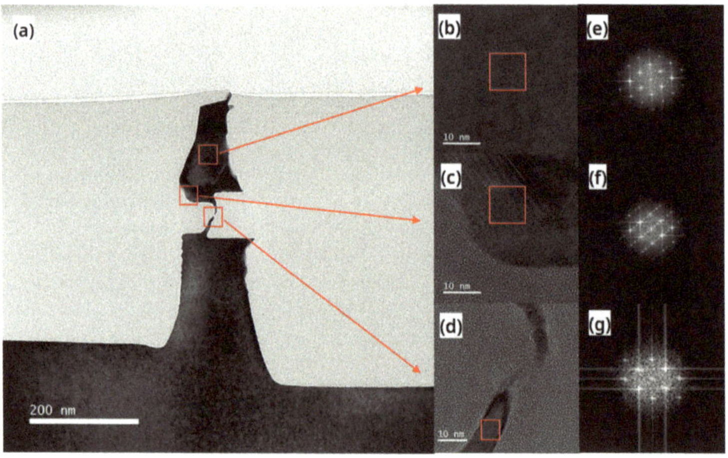

Figure 7. (**a**) TEM image, (**b–d**) HRTEM images, and (**e–g**) FFT images of the cross-section of the device after laser annealing process.

3.3. Characterization and Materials Analysis of the Device

After completing all the process steps, FIB-TEM analysis and electrical performance testing were performed on a RC-VCNFET. Figure 8 shows the high-angle annular dark-field (HAADF) image and EDS mapping of the channel region of the device with HKMG. The Si in black contrast in Figure 8a shows the integrity of the RC-VCNFET device's channel, and the channel's minimum thickness was measured to be 3.8 nm. Figure 8b indicates that the thickness of the RC-channel is less than the limited EDS spatial resolution. That means the channel is very thin, so that the device will have a strong gate control capability. Figure 8c,d show that HKMG deposited by ALD has an excellent conformal filling ability for tens-of-nanometer-scale C-shaped cavities. Moreover, the gate length of the RC-VCNFET is about 49 nm, which is defined by the thickness of the tungsten filled in the gap on the left side of the RC-VCNFET.

Figure 9a shows the top view TEM image of the device at the channel height, and the design size of the mandrel layer is 1 μm × 1 μm. The white contrast Si in Figure 9a shows the continuity of the device channel in the plane direction. Due to the diffraction effect of light, the 1 μm × 1 μm rectangular mandrel is significantly distorted at the corners, making the square mandrel on the original layout an approximately circular structure. The actual channel width of this RC-VCNFET is measured to be about 2.76 μm, and its original design value on the layout is 4 μm. Figure 9b,c also show the continuity of the RC-VCNFET device channel with an average thickness of about 7.74 nm. Figure 9d,e also prove that the ALD-grown HfO_2 and tungsten films have perfect conformality at the nanoscale.

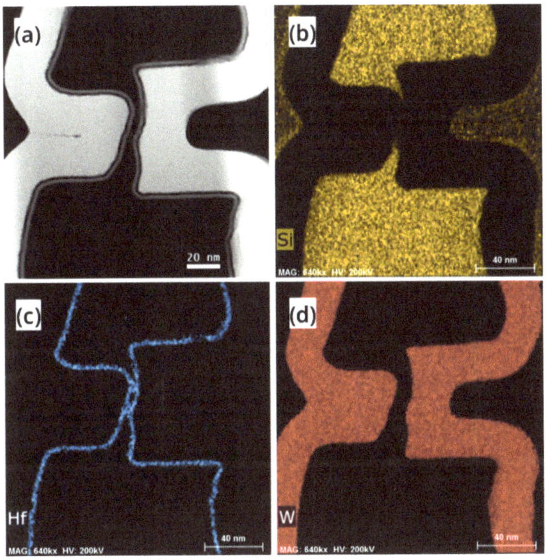

Figure 8. (**a**) HAADF image, (**b**) Si element, (**c**) Hf element, (**d**) W element mapping of the cross-section of the device.

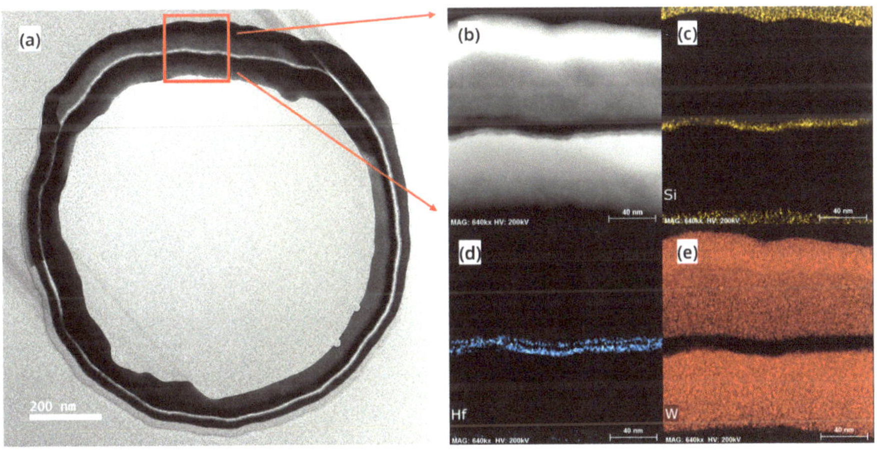

Figure 9. (**a**) TEM image, (**b**) HAADF image, (**c**) Si element, (**d**) Hf element, (**e**) W element mapping of the top view of the device.

3.4. Electrical Properties of the Device

Figure 10a,b show the Id-Vg and Id-Vd curves of the RC-VCNFET device, respectively, where I_{on} is 15.7 μA/μm (I_D @ $V_{OV} = V_G - V_T = 1$ V, $V_{DS} = 1.25$ V), I_{off} is 5.48×10^{-6} pA/μm (I_D @ $V_{OV} = V_G - V_T = -0.50$ V, $V_{DS} = 1.25$ V), threshold voltage V_T is 0.75 V, subthreshold swing (SS) is 62.91 mV/dec, drain-induced barrier lowering (DIBL) is 6.16 mV/V, and the on–off ratio is 2.86×10^6. The above electrical parameters show that the current driving performance of the current device is relatively low. However, the device's off-state leakage current and gate control capability are excellent, which can be attributed to the fact that the thickness of the RC-channel is only 3.8 nm. It should be noted that these electrical characteristics only show the driving performance of the RC-

VCNFET fabricated in the early stage. The device's performance can be further improved by optimizing the laser annealing recrystallization process, source/drain contact, and other process modules.

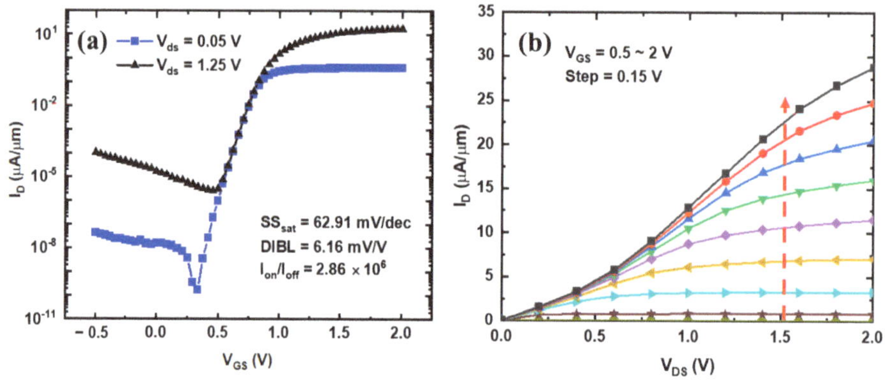

Figure 10. (**a**) Transfer characteristic curve and (**b**) output characteristic curve of the device.

4. Conclusions

The RC-VCNFETs with a self-aligned structure were proposed and fabricated in this work. Its related process modules were developed: selective etching of SiN cavity, channel Si cap growth, and recrystallization channel formation. The RC-VCNFET was characterized by TEM, EDS, and other analysis methods. Finally, the electrical characteristics of the device were tested and analyzed. The device showed superior gate control capability, with a SS of 62.91 mV/dec and a DIBL of 6.16 mV/V.

Author Contributions: Conceptualization, Z.C. and H.Z.; Data curation, Z.C.; Funding acquisition, H.Z.; Project administration, H.Z.; Supervision, H.Z. and G.W.; Writing—original draft, Z.C.; Writing—review & editing, Z.C., H.Z. and G.W.; Device process integration, Z.C., Q.W., Z.X., Y.Z. (Yongkui Zhang), S.L., Y.D., Z.Y. and Y.Z. (Yantong Zheng); Laser annealing process tuning, J.L.; RTCVD process tuning, W.X.; Pre-clean process and ALD process tuning, Z.K.; TEM measurement, A.D.; AFM measurement, J.Y. All authors have read and agreed to the published version of the manuscript.

Funding: This work was supported by the National Natural Science Foundation of China (Grant No. 92064002) and the Beijing Superstring Academy of Memory Technology (No. E2DF06X003).

Data Availability Statement: The data presented in this study are available on request from the corresponding author.

Conflicts of Interest: The authors declare no conflict of interest.

References

1. Chang, C.-H.; Chang, V.S.; Pan, K.H.; Lai, K.T.; Lu, J.H.; Ng, J.A.; Chen, C.Y.; Wu, B.F.; Lin, C.J.; Liang, C.S. Critical Process Features Enabling Aggressive Contacted Gate Pitch Scaling for 3 nm CMOS Technology and Beyond. In Proceedings of the 2022 International Electron Devices Meeting (IEDM), San Francisco, CA, USA, 3–7 December 2022; IEEE: Piscataway, NJ, USA, 2022; pp. 27.1.1–27.1.4.
2. Lee, S.-Y.; Kim, S.-M.; Yoon, E.-J.; Oh, C.-W.; Chung, I.; Park, D.; Kim, K. A Novel Multibridge-Channel MOSFET (MBCFET): Fabrication Technologies and Characteristics. *IEEE Trans. Nanotechnol.* **2003**, *2*, 253–257. [CrossRef]
3. Bae, G.; Bae, D.-I.; Kang, M.; Hwang, S.M.; Kim, S.S.; Seo, B.; Kwon, T.Y.; Lee, T.J.; Moon, C.; Choi, Y.M. 3nm GAA Technology Featuring Multi-Bridge-Channel FET for Low Power and High Performance Applications. In Proceedings of the 2018 IEEE International Electron Devices Meeting (IEDM), San Francisco, CA, USA, 1–5 December 2018; IEEE: Piscataway, NJ, USA, 2018; Session 28, No. 7.
4. Gupta, A.; Pedreira, O.V.; Arutchelvan, G.; Zahedmanesh, H.; Devriendt, K.; Mertens, H.; Tao, Z.; Ritzenthaler, R.; Wang, S.; Radisic, D. Buried Power Rail Integration with FinFETs for Ultimate CMOS Scaling. *IEEE Trans. Electron Devices* **2020**, *67*, 5349–5354. [CrossRef]

5. Gupta, A.; Mertens, H.; Tao, Z.; Demuynck, S.; Bömmels, J.; Arutchelvan, G.; Devriendt, K.; Pedreira, O.V.; Ritzenthaler, R.; Wang, S. Buried Power Rail Integration with Si FinFETs for CMOS Scaling beyond the 5 nm Node. In Proceedings of the 2020 IEEE Symposium on VLSI Technology, Honolulu, HI, USA, 16–19 June 2020; IEEE: Piscataway, NJ, USA, 2020; pp. 1–2.
6. Gupta, A.; Pedreira, O.V.; Tao, Z.; Mertens, H.; Radisic, D.; Jourdan, N.; Devriendt, K.; Heylen, N.; Wang, S.; Chehab, B. Buried Power Rail Scaling and Metal Assessment for the 3 nm Node and Beyond. In Proceedings of the 2020 IEEE International Electron Devices Meeting (IEDM), San Francisco, CA, USA, 12–18 December 2020; IEEE: Piscataway, NJ, USA, 2020; pp. 20–23.
7. Prasad, D.; Nibhanupudi, S.T.; Das, S.; Zografos, O.; Chehab, B.; Sarkar, S.; Baert, R.; Robinson, A.; Gupta, A.; Spessot, A. Buried Power Rails and Back-Side Power Grids: Arm® CPU Power Delivery Network Design beyond 5nm. In Proceedings of the 2019 IEEE International Electron Devices Meeting (IEDM), San Francisco, CA, USA, 7–11 December 2019; IEEE: Piscataway, NJ, USA, 2019; Session 19, No. 1.
8. Hossen, M.O.; Chava, B.; Van der Plas, G.; Beyne, E.; Bakir, M.S. Power Delivery Network (PDN) Modeling for Backside-PDN Configurations With Buried Power Rails and μTSVs. *IEEE Trans. Electron Devices* **2019**, *67*, 11–17. [CrossRef]
9. Song, K.-W.; Kim, J.-Y.; Yoon, J.-M.; Kim, S.; Kim, H.; Chung, H.-W.; Kim, H.; Kim, K.; Park, H.-W.; Kang, H.C. A 31 Ns Random Cycle VCAT-Based 4F ^{2} DRAM With Manufacturability and Enhanced Cell Efficiency. *IEEE J. Solid-State Circuits* **2010**, *45*, 880–888. [CrossRef]
10. Lee, S.-W.; Kim, S.-Y.; Hwang, K.-M.; Jin, I.K.; Hur, J.; Kim, D.-H.; Son, J.W.; Kim, W.-K.; Choi, Y.-K. A Comprehensive Study of a Single-Transistor Latch in Vertical Pillar-Type FETs with Asymmetric Source and Drain. *IEEE Trans. Electron Devices* **2018**, *65*, 5208–5212. [CrossRef]
11. Cho, Y.; Choi, P.; Hyeon, Y.; Song, J.; Hwang, Y.; Choi, B. Novel Band-to-Band Tunneling Body Contact (BTBC) Structure to Suppress the Floating-Body Effect in a Vertical-Cell DRAM. *IEEE Electron Device Lett.* **2018**, *39*, 1860–1863. [CrossRef]
12. Cho, Y.S.; Choi, P.H.; Kim, K.H.; Park, J.M.; Hwang, Y.S.; Hong, H.S.; Lee, K.P.; Choi, B.D. Stretched Tunnelling Body Contact Structure for Suppressing the FBE in a Vertical Cell DRAM. *Electron. Lett.* **2019**, *55*, 1252–1253. [CrossRef]
13. Cho, Y.; Kim, H.; Jung, K.; Kim, B.; Hwang, Y.; Hong, H.; Choi, B. Suppression of the Floating-Body Effect of Vertical-Cell DRAM with the Buried Body Engineering Method. *IEEE Trans. Electron Devices* **2018**, *65*, 3237–3242. [CrossRef]
14. Schmidt, V.; Riel, H.; Senz, S.; Karg, S.; Riess, W.; Gösele, U. Realization of a Silicon Nanowire Vertical Surround-Gate Field-Effect Transistor. *Small* **2006**, *2*, 85–88. [CrossRef] [PubMed]
15. Goldberger, J.; Hochbaum, A.I.; Fan, R.; Yang, P. Silicon Vertically Integrated Nanowire Field Effect Transistors. *Nano Lett.* **2006**, *6*, 973–977. [CrossRef]
16. Jagannathan, H.; Anderson, B.; Sohn, C.-W.; Tsutsui, G.; Strane, J.; Xie, R.; Fan, S.; Kim, K.I.; Song, S.; Sieg, S. Vertical-Transport Nanosheet Technology for CMOS Scaling beyond Lateral-Transport Devices. In Proceedings of the 2021 IEEE International Electron Devices Meeting (IEDM), San Francisco, CA, USA, 11–16 December 2021; IEEE: Piscataway, NJ, USA, 2021; Session 26, No. 1.
17. Tsutsui, G.; Song, S.; Strane, J.; Xie, R.; Qin, L.; Zhang, C.; Schmidt, D.; Fan, S.; Hong, B.; Jung, Y. *Hardware Based Performance Assessment of Vertical-Transport Nanosheet Technology*; IEEE International Electron Devices Meeting: San Francisco CA, USA, 2022.
18. Zhang, Y.; Ai, X.; Yin, X.; Zhu, H.; Yang, H.; Wang, G.L.; Li, J.J.; Du, A.Y.; Li, C.; Huang, W.X. Vertical Sandwich GAA FETs with Self-Aligned High-k Metal Gate Made by Quasi Atomic Layer Etching Process. *IEEE Trans. Electron Devices* **2021**, *68*, 2604–2610. [CrossRef]
19. Yin, X.; Zhang, Y.; Zhu, H.; Wang, G.L.; Li, J.J.; Du, A.Y.; Li, C.; Zhao, L.H.; Huang, W.X.; Yang, H. Vertical Sandwich Gate-All-around Field-Effect Transistors with Self-Aligned High-k Metal Gates and Small Effective-Gate-Length Variation. *IEEE Electron Device Lett.* **2019**, *41*, 8–11. [CrossRef]
20. Li, C.; Zhu, H.; Zhang, Y.; Yin, X.; Jia, K.; Li, J.; Wang, G.; Kong, Z.; Du, A.; Yang, T. Selective Digital Etching of Silicon–Germanium Using Nitric and Hydrofluoric Acids. *ACS Appl. Mater. Interfaces* **2020**, *12*, 48170–48178. [CrossRef] [PubMed]
21. Li, C.; Zhu, H.; Zhang, Y.; Wang, Q.; Yin, X.; Li, J.; Wang, G.; Kong, Z.; Ai, X.; Xie, L. First Demonstration of Novel Vertical Gate-All-around Field-Effect-Transistors Featured by Self-Aligned and Replaced High-κ Metal Gates. *Nano Lett.* **2021**, *21*, 4730–4737. [CrossRef] [PubMed]
22. Xiao, Z.R.; Zhu, H.L.; Wang, Q.; Chen, Z.; Liu, Z.Y.; Zhang, Y.K.; Yan, Z.J.; Shi, Y.F.; Zhou, N.; Li, J.J.; et al. Vertical N-Type and P-Type Nanosheet FETs With C-Shaped Channel. *IEEE Trans. Electron Devices* **2023**, *70*, 1380–1385. [CrossRef]
23. Xiao, Z.R.; Wang, Q.; Zhu, H.L.; Chen, Z.; Zhang, Y.K.; Li, J.J.; Zhou, N.; Gao, J.F.; Ai, X.Z.; Lu, S.S.; et al. Vertical C-Shaped-Channel Nanosheet FETs Featured With Precise Control of Both Channel-Thickness and Gate-Length. *IEEE Electron Device Lett.* **2022**, *43*, 1183–1186. [CrossRef]

nanomaterials

MDPI

Article

Low-Temperature (≤500 °C) Complementary Schottky Source/Drain FinFETs for 3D Sequential Integration

Shujuan Mao [1], Jianfeng Gao [1], Xiaobin He [1], Weibing Liu [1], Jinbiao Liu [1], Guilei Wang [2], Na Zhou [1], Yanna Luo [1,3], Lei Cao [1,3], Ran Zhang [1], Haochen Liu [1], Xun Li [1], Yongliang Li [1], Zhenhua Wu [1], Junfeng Li [1], Jun Luo [1,3], Chao Zhao [2], Wenwu Wang [1,3] and Huaxiang Yin [1,3,*]

[1] Institute of Microelectronics, Chinese Academy of Sciences, Beijing 100029, China; maoshujuan@ime.ac.cn (S.M.); gaojianfeng@ime.ac.cn (J.G.); hexiaobin@ime.ac.cn (X.H.); liuweibing@ime.ac.cn (W.L.); liujinbiao@ime.ac.cn (J.L.); zhouna@ime.ac.cn (N.Z.); luoyanna@ime.ac.cn (Y.L.); caolei@ime.ac.cn (L.C.); zhangran@ime.ac.cn (R.Z.); liuhaochen@ime.ac.cn (H.L.); lixun@ime.ac.cn (X.L.); liyongliang@ime.ac.cn (Y.L.); wuzhenhua@ime.ac.cn (Z.W.); lijunfeng@ime.ac.cn (J.L.); luojun@ime.ac.cn (J.L.); wangwenwu@ime.ac.cn (W.W.)

[2] Beijing Superstring Academy of Memory Technology, Beijing 100176, China; guilei.wang@bjsamt.org.cn (G.W.); chao.zhao@bjsamt.org.cn (C.Z.)

[3] School of Microelectronics, University of Chinese Academy of Sciences, Beijing 100049, China

* Correspondence: yinhuaxiang@ime.ac.cn

Abstract: In this work, low-temperature Schottky source/drain (S/D) MOSFETs are investigated as the top-tier devices for 3D sequential integration. Complementary Schottky S/D FinFETs are successfully fabricated with a maximum processing temperature of 500 °C. Through source/drain extension (SDE) engineering, competitive driving capability and switching properties are achieved in comparison to the conventional devices fabricated with a standard high-temperature (≥1000 °C) process flow. Schottky S/D PMOS exhibits an ON-state current (I_{ON}) of 76.07 μA/μm and ON-state to OFF-state current ratio (I_{ON}/I_{OFF}) of 7×10^5, and those for NMOS are 48.57 μA/μm and 1×10^6. The CMOS inverter shows a voltage gain of 18V/V, a noise margin for high (NM_H) of 0.17 V and for low (NM_L) of 0.43 V, with power consumption less than 0.9 μW at V_{DD} of 0.8 V. Full functionality of CMOS ring oscillators (RO) are further demonstrated.

Keywords: 3D sequential integration; low thermal budget; Schottky S/D FinFETs; inverter

1. Introduction

The technology of 2D planar scaling is now facing major limitations, and in order to extend the semiconductor roadmap, 3D sequential integration, which consists of stacking transistors on top of each other, has been envisioned [1,2]. As its name suggests, transistor layers are processed sequentially, i.e., the top tier is processed and stacked above the already fabricated bottom tier in 3D sequential integration. This technology can enhance circuit density and functionality without the requirement of further reduction in device dimensions. To maintain the integrity of what is below, namely the bottom devices, interconnections and bonding interface, the thermal budget for top-tier fabrication is required to be no more than 550 °C [3–5].

However, source/drain (S/D) activation is typically performed by spike annealing at high temperature (≥1000 °C). Decreasing thermal budget will impair CMOS device performance. Much work has been undertaken to circumvent the thermal budget limitation. For instance, nanosecond laser annealing (NLA) [6] and solid-phase epitaxial regrowth (SPER) [7] were used to activate S/D as the alternatives to high-temperature spike annealing and low-temperature materials, such as poly-Si [8,9], Ge [10,11], III-V [12,13] and transparent amorphous oxide [14,15] were implemented to replace monocrystalline Si as the channel of top-tier devices. Particularly interesting is the exploration of junctionless MOSFETs as the top-tier devices with the elimination of S/D activation [16,17]. Even though impressive device performances have been achieved with such approaches, there remain several issues. NLA and SPER often incur high process cost and low throughput, and low-temperature materials are not fully compatible with current Si technology, leading to a risk of poor yield at a large scale. Additionally, junctionless devices need an extra-high-temperature ($\geq$1000 °C) annealing to activate the channel before the top silicon layer transfer, which is likely to induce mobility degradation and threshold voltage (V_{TH}) variation. Hence, low-temperature devices based on Si technology may be further developed.

Schottky S/D MOSFETs, using metal silicide to replace doped silicon as S/D [18], hold an inherent superiority in process thermal budget over the conventional and junctionless devices, with no need of a standard high-temperature annealing to activate S/D and channel. Therefore, in this work, low-temperature complementary Schottky S/D FinFETs are processed at a temperature as low as 500 °C, and the electrical characteristics are investigated to evaluate the feasibility of being used as the top-tier logic devices in 3D sequential technology. To our knowledge, previous investigations on Schottky S/D devices have been mostly focused on short-channel effects, with the fabrication thermal budget never lower than 600 °C [19–24]. This is the first demonstration of complementary S/D FinFETs with all process steps below 500 °C toward 3D sequential integration.

2. Device Fabrication

The process flow of Schottky S/D FinFETs is summarized in Figure 1a. SOI wafers measuring 200 mm with top Si of 40 nm and BOX of 145 nm were used as the starting materials to mimic the bonded substrate of top-tier devices. The replacement metal gate (RMG) process was adopted, and all process steps were set below the typical thermal budget of 550 °C for compatibility with 3D sequential integration [3–5]. According to the principle of Schottky S/D MOSFETs [18], the electrical property is primarily determined by the Schottky junction barrier between S/D and channel. In order to realize high performance, a doped source/drain extension (SDE) to lower the Schottky junction barrier, illustrated in Figure 1b, was first explored for pFinFETs by two methods, i.e., SDE first (SDE^{1st}) and SDE last (SDE^{last}). In the SDE^{1st} scheme, SDE implantation was performed before the spacer and followed by silicidation annealing, and SDE was formed by dopant segregation at the silicide/Si interface during silicidation. In contrast, SDE implantation was performed after silicide formation in the SDE^{last} scheme, and an additional rapid thermal annealing (RTA), also named drive-in annealing, was used to drive the dopant to segregate at the silicide/Si interface, forming SDE. An amount of 3 nm Ni was deposited by sputtering, and B 1.5 keV 2 $\times$ 10^{15} cm^{-2} was implemented for SDE implantation in both schemes. A split, shown in Table 1, was further performed to investigate the impact of the SDE process thermal budget on device performance. Afterwards, the process flow of complementary Schottky S/D FinFETs was developed with optimal SDE engineering. Gate stacks of HfO_2/TiN/TaN and HfO_2/TiAl/TiN were separately applied to pFinFETs and nFinFETs for V_{TH} adjustment. The fabrication was completed with tungsten contact plug and Al metallization. Conventional pFinFETs, with and without silicide, were also fabricated with a standard high-temperature ($\geq$1000 °C) process flow (Figure 1c) for comparison. It is worth noting that 8-inch industrial equipment was used for the fabrication in our experiment, and the process uniformity of within wafer, wafer-to-wafer and lot-to-lot was controllable and reproducible. Current-voltage measurements (112 measurement sites for each wafer) were performed using a HP4156 parameter analyser.

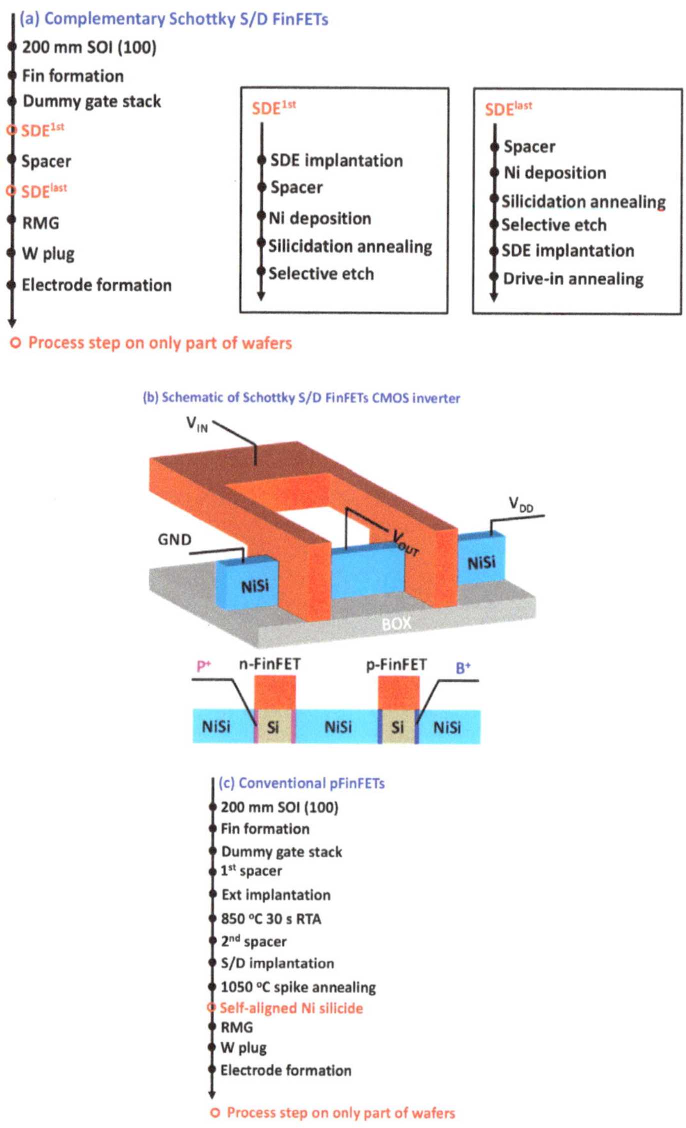

Figure 1. (**a**) Process flow of Schottky S/D FinFETs in this work, (**b**) Schematic layout and cross section of the complementary Schottky S/D FinFETs inverter, (**c**) Standard high-temperature process flow of conventional device.

Table 1. Split of thermal budget for SDE formation.

Item [1]	310 °C SDE 1st	450 °C SDE last	550 °C SDE last
Silicidation annealing	310 °C 60 s	310 °C 60 s	500 °C 30 s
Drive-in annealing	NA	450 °C 60 s	550 °C 60 s

[1] All annealing steps were performed by RTA.

3. Results and Discussion

3.1. SDE Engineering

To develop the optimal SDE process, SDE engineering, shown in Table 1, was performed on pFinFETs. The I_{DS}-V_{GS} characteristics of the Schottky S/D pFinFETs with three SDE processes are noted in Figure 2a–c. Decent switching properties are demonstrated for all three, with ON-state to OFF-state current ratios (I_{ON}/I_{OFF}) of 5 orders, and excellent swings (SSs) of around 61 mV/decade obtained for physical gate length (L_G) of 500 nm. As compared in Figure 2d, the top I_{ON} was achieved with 550 °C SDE^{last}, followed by 310 °C SDE^{1st} and then 450 °C SDE^{last}, with the values of 55.49 μA/μm, 40.78 μA/μm and 25.23 μA/μm at V_{DS} = −0.8 V; whereas larger I_{OFF} of 0.39 nA/μm and 0.21 nA/μm were resolved for 310 °C SDE^{1st} and 450 °C SDE^{last} with respect to that of 0.08 nA/μm for 550 °C SDE^{last}. Further, the V_{TH} defined as V_{GS} corresponding to $I_{DS} = 10^{-7}$ A/μm were −0.13 V, −0.16 V and −0.16 V for the devices with 310 °C SDE^{1st}, 450 °C SDE^{last} and 550 °C SDE^{last} processes, respectively. A reduction of around 30 mV in V_{TH} was demonstrated for 310 °C SDE^{1st}.

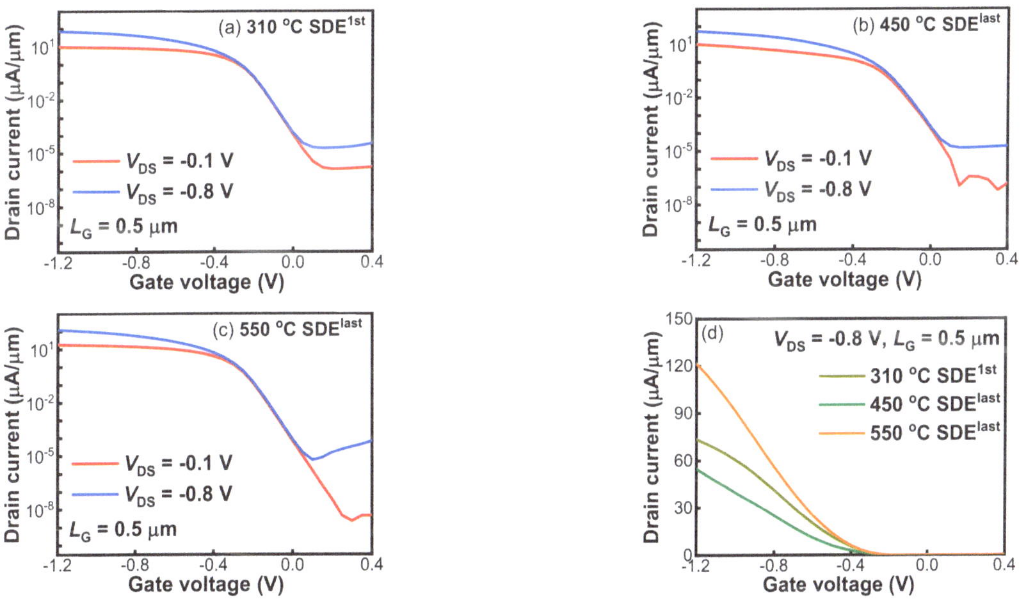

Figure 2. I_{DS}-V_{GS} characteristics of the Schottky S/D pFinFETs with (**a**) 310 °C SDE^{1st}, (**b**) 450 °C SDE^{last} and (**c**) 550 °C SDE^{last} processes; (**d**) Comparison of I_{DS} at V_{DS} = −0.8 V between the three.

Figure 3 shows the S/D series resistance ($R_{S/D}$) extracted by fitting ON-state resistance (R_{ON}) vs. $1/(V_{GS}$-$V_{TH})$ at high gate bias in linear mode [25]. The $R_{S/D}$ measurements at V_{GS} = −0.8 V were about 6.31 kΩ/μm, 8.71 kΩ/μm and 2.46 kΩ/μm for the devices with 310 °C SDE^{1st}, 450 °C SDE^{last} and 550 °C SDE^{last} processes, respectively. Such high $R_{S/D}$ values are ascribed to the nano-fins scheme and to insufficiently silicided S/D, as uncovered in Figure 4. Due to the ultrathin Ni of 3 nm, about 16 nm of silicide was formed, and yet around 24 nm silicon remained unsilicided. Figure 5a further shows the I_{ON} dependence on the SDE process thermal budget, which is involved with the annealing steps in Table 1. It should be noted that the heating-up and cooling-down periods of an annealing process are ignored when defining a thermal budget. It was revealed that lowering the SDE process thermal budget degraded I_{ON} in the SDE^{last} scheme; it seems that SDE^{1st} prevails over SDE^{last} in driving capability at a lower thermal budget level.

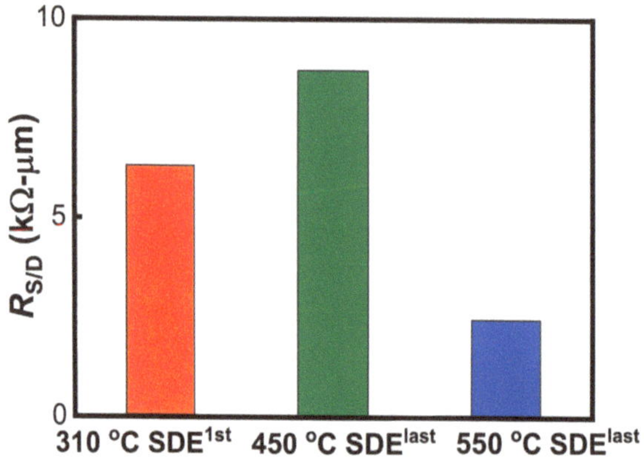

Figure 3. $R_{S/D}$ at $V_{DS} = -0.1$ V of the Schottky S/D pFinFETs with different SDE processes.

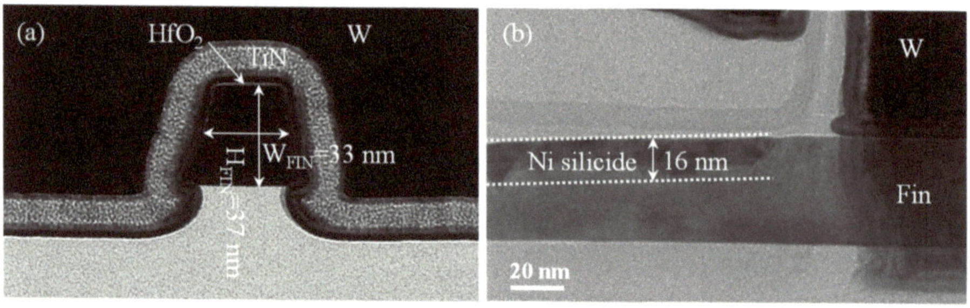

Figure 4. Cross-sectional transmission electron microscope (XTEM) images of the Schottky S/D pFinFETs (**a**) across and (**b**) along fins with gate stack covering, fabricated with 550 °C SED^{last} process.

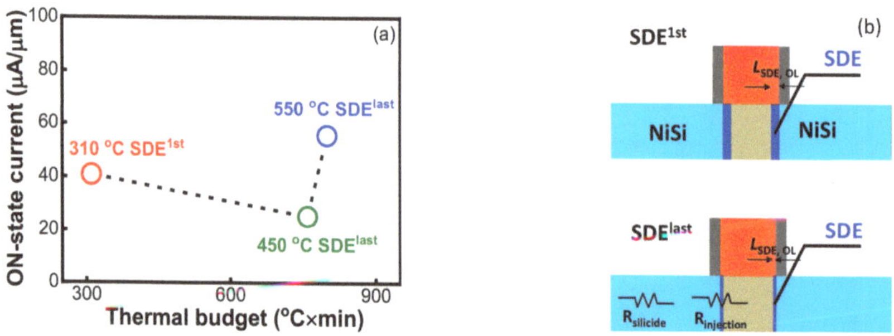

Figure 5. (**a**) I_{ON} dependence on SDE process thermal budget for Schottky S/D pFinFETs, (**b**) Schematic of gate-to-SDE overlap ($L_{SDE,\,OL}$) and $R_{S/D}$ component for the devices with SDE^{1st} and SDE^{last} processes.

Correlating the electrical results with SDE engineering, two main findings can be made. First, SDE^{1st} holds the advantage in I_{ON} at a lower thermal budget level, which results from the larger gate-to-SDE overlap ($L_{SDE,\,OL}$), by performing SDE implantation before the spacer, as illustrated in Figure 5b. The reduced V_{TH} of -0.13 V, larger I_{OFF} of 0.39 nA/μm and lower $R_{S/D}$ of 6.31 kΩ/μm for 310 °C SDE^{1st}, with reference to 450 °C SDE^{last}, justify this point, which was performed with the same silicidation annealing. Second, in the SDE^{last} scheme, increasing the thermal budget will lower $R_{S/D}$, and thus improve I_{ON}. The lowered $R_{S/D}$ is probably attributable to the reductions in silicide resistance and injection resistance (Figure 5b). It is known that an Ni/Si solid state reaction forms high-resistance Ni_2Si at 250–400 °C and low-resistance NiSi at 400–700 °C [26]. With the raising of silicidation annealing from 310 °C to 500 °C for 450 °C SDE^{last} and 550 °C SDE^{last} (Table 1), the sheet resistance of Ni silicide decreases from around 457 Ω to 123 Ω, measured with a four-point probe system. Additionally, it has been demonstrated that increasing drive-in annealing will boost dopant segregation at the silicide/Si interface, leading to an enhanced Schottky junction barrier lowering [27–29]. Since the injection resistance is proportional to the Schottky junction barrier, its reduction can be expected for 550 °C SDE^{last} with respect to 450 °C SDE^{last}, with drive-in annealing at 550 °C for the former and 450 °C for the later. One may argue that the mobility could differ with SDE process thermal budget, affecting device performance. Since no channel doping was performed for Schottky devices and all samples were tested at 300 K, it is supposed that the mobility was almost the same and the difference in I_{DS}-V_{GS} characteristics in Figure 2 was primarily caused by $R_{S/D}$ and V_{TH}.

3.2. Low-Temperature Schottky S/D FinFETs vs. Conventional High-Temperature FinFETs

In accordance with the findings in Section 3.1, the fabrication of complementary Schottky S/D FinFETs with optimal SDE engineering was developed toward 3D sequential integration. The thickness of Ni was increased from 3 nm to 6 nm and 5% Pt. was added, and a two-step RTA (310 °C 60 s + selective etch + 500 °C 10 s) method was explored for silicide formation so as to avoid abnormal Ni diffusion [30,31]. Meanwhile, the drive-in annealing was further reduced from 550 °C 60 s to 500 °C 60 s for better compatibility with 3D sequential integration. Figure 6 shows the I_{DS}-V_{GS} characteristics of the complementary Schottky S/D FinFETs with L_G = 500 nm. The I_{ON} and I_{ON}/I_{OFF} ratios for pFinFETs were 76.07 µA/µm and 7 × 10^5, and those for nFinFETs were 48.57 and 1 × 10^6 µA/µm, at V_{DS} = ±0.8 V. The corresponding V_{TH} values were around −0.16 V and 0.3 V. Clearly, with S/D fully silicided as confirmed by XTEM images (not shown), an improvement in I_{ON} was achieved for pFinFETs due to the reduced $R_{S/D}$.

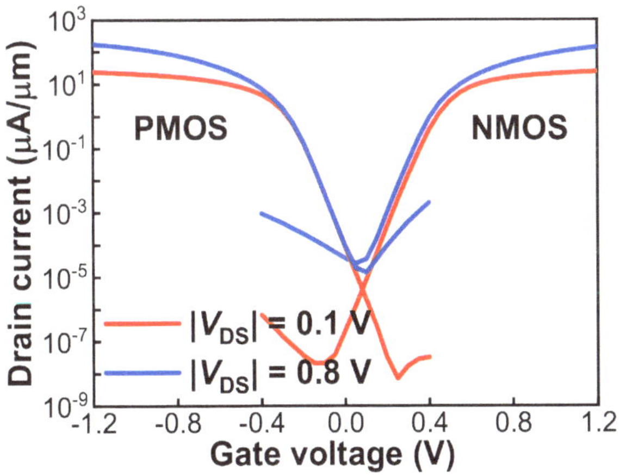

Figure 6. I_{DS}-V_{GS} characteristics of low-temperature complementary Schottky S/D FinFETs with L_G = 500 nm.

A comparison was made between low-temperature Schottky S/D pFinFETs and conventional pFinFETs fabricated with a standard high-temperature (≥ 1000 °C) process flow (Figure 1c). As shown in Figure 7a, with silicide formation, the I_{ON} of conventional pFinFETs is significantly improved. Since no shift of V_{TH} was evidenced with silicide (Figure 7c), the I_{ON} improvement is primarily attributable to the decrease in $R_{S/D}$ from 29 kΩ/µm to

2 kΩ/μm (Figure 7b). The low-temperature Schottky S/D pFinFETs exhibited higher I_{ON} than the conventional high-temperature devices, whether they were with (w/) or without (w/o) silicide, owing to the lower $R_{S/D}$ (Figure 7b) and V_{TH} (Figure 7c). As compared to the conventional device, the Schottky S/D device using silicide as S/D holds an inherent advantage in $R_{S/D}$, and its fabrication is fully compatible with current Si technology; competitive performance can be obtained with a 500 °C drive-in annealing to form SDE; moreover, no annealing at ≥1000 °C is needed to activate the channel with respect to a junctionless device [16,17]. Hence, it is indeed feasible to adopt Schottky S/D FinFETs as the top-tier devices in 3D sequential technology.

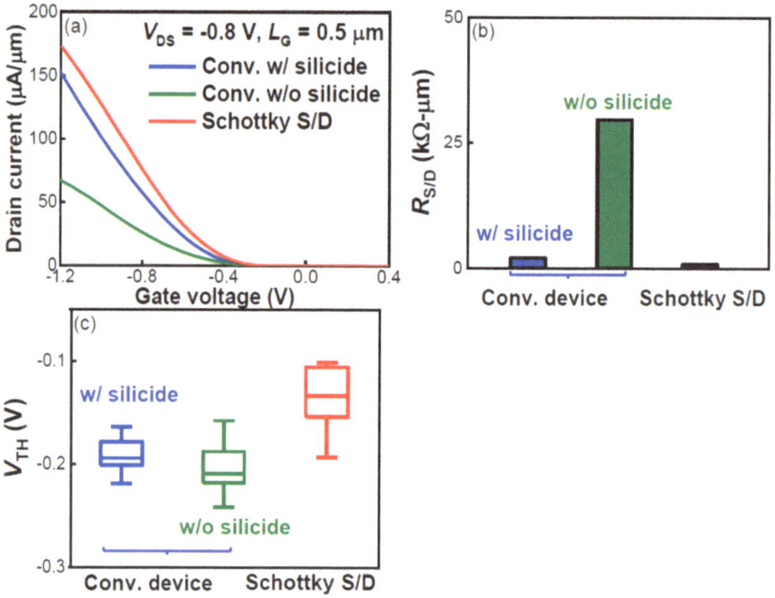

Figure 7. Comparison of (**a**) I_{DS} at V_{DS} = −0.8 V, (**b**) $R_{S/D}$ and (**c**) V_{TH} between low-temperature Schottky S/D device and conventional (conv.) high-temperature device.

3.3. Inverter Characterization

The CMOS inverter voltage transfer characteristics (VTC) at V_{DD} ranging from 0.3 V to 1 V by step of 0.1 V are presented in Figure 8a. The source of Schottky S/D NMOS was connected to the ground potential, while the source of Schottky S/D PMOS was attached to V_{DD}. Both transistors shared the silicided drain contact forming the output terminal of inverter V_{OUT}, as illustrated in Figure 1b. Well-behaved VTC was obtained, with a low-to-high output dynamic that reached rail-to-rail supply voltage range. This indicates that the subthreshold leakage currents of both transistors were sufficiently low to not degrade high and low logic states. It is noted that the transition of the inverter VTC was not located at $V_{DD}/2$, due to the uncompensated asymmetry of V_{TH} between pFinFETs and nFinFETs (Figure 6). The transition of the inverter VTC was shifted by the same amount of about 0.14 V. A gate metal work function adjustment could be applied to optimize V_{TH} symmetry to further improve the inverter VTC. Almost a constant voltage gain ($\Delta V_{OUT}/\Delta V_{IN}$) of 18 *v/v* was achieved at V_{DD} in the range of 0.3 V ~ 0.8 V (Figure 8b), suggesting a great potential of our inverter in low-power and high-performance 3D sequential integration. In order to estimate the noise margin (*NM*), a piecewise approximation of the VTC was used here to determine the boundary of the transition zone. As illustrated in Figure 9a, the output voltage and input voltage for high (V_{OH}, V_{IH}) as well as for low (V_{OL}, V_{IL}) were defined by the position of points where $dV_{OUT}/dV_{IN} = -1$. *NM* for high input

($NM_H = V_{OH} - V_{IH}$) of 0.17 V and *NM* for low input ($NM_L = V_{IL} - V_{OL}$) of 0.43 V at V_{DD} = 0.8 V were obtained. Figure 9b shows the static power consumption as a function of V_{IN} at V_{DD} = 0.8 V. The maximum static power consumption for V_{IN} sweeping from 0 V to 0.8 V at V_{DD} = 0.8 V was less than 0.9 μW. In Figure 9c, CMOS ring oscillators (RO) composed by 101 stages were successfully operated in low-temperature Schottky S/D FinFETs. Again, this validates the feasibility of Schottky S/D FinFETs as the top-tier devices in 3D sequential technology.

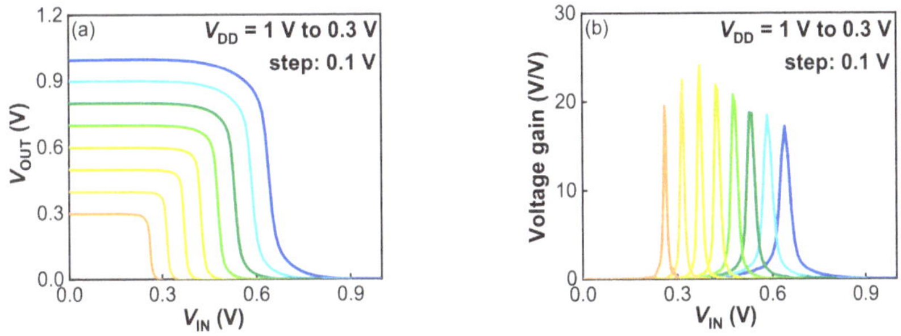

Figure 8. (**a**) Inverter VTC at V_{DD} ranging from 1 V down to 0.3 V, (**b**) Corresponding voltage gains.

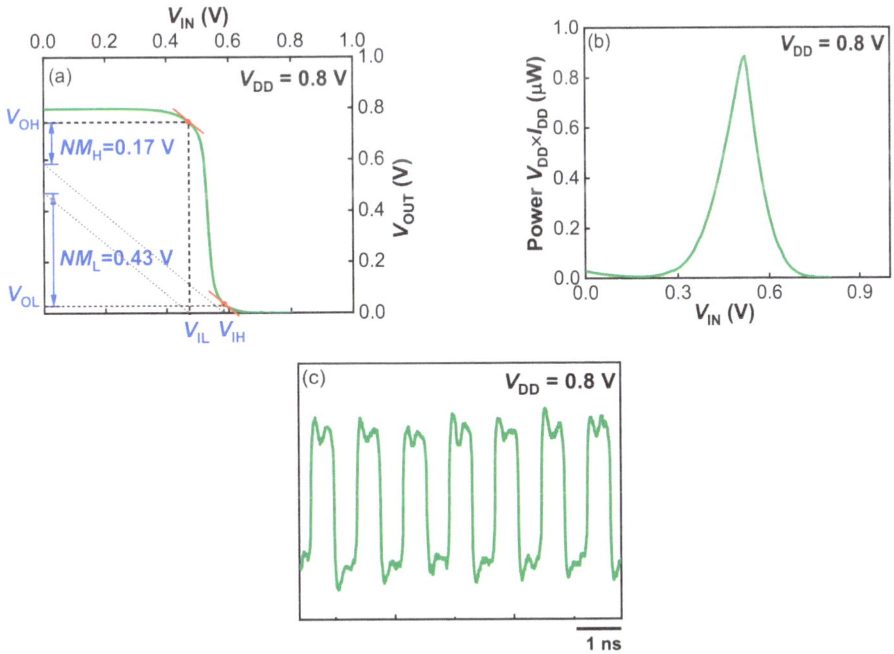

Figure 9. (**a**) Linear approximation of the VTC to estimate the static NMs, (**b**) Static power consumption with respect to V_{IN}, (**c**) Characteristics of 101-stage CMOS RO based on low-temperature Schottky S/D FinFETs.

4. Conclusions

In conclusion, low-temperature complementary Schottky S/D FinFETs were proposed as the top-tier devices for 3D sequential integration and were experimentally demonstrated in this work. The thermal budget for fabrication was no more than 500 °C. and the entire process flow was fully compatible with current Si technology. With optimal SDE engineering and competitive I_{ON} values of 76.07 μA/μm and 48.57 μA/μm, I_{ON}/I_{OFF} ratios of 7×10^5 and 1×10^6 at V_{DD} = 0.8 V were obtained for pFinFETs and nFinFETs, respectively. Excellent CMOS inverter and functional CMOS RO are successfully explored, offering a new method of high-performance 3D VLSI CMOS integration.

Author Contributions: Conceptualization, methodology and writing—original draft preparation, S.M. and J.G.; methodology, validation and formal analysis, X.H., W.L., J.L. (Jinbiao Liu) and G.W.; investigation and data curation, N.Z., Y.L. (Yanna Luo) and L.C.; resources, R.Z., H.L. and X.L.; writing—review and editing, Y.L. (Yongliang Li) and Z.W.; supervision, project administration, J.L. (Junfeng Li), J.L. (Jun Luo), C.Z., W.W. and H.Y. All authors have read and agreed to the published version of the manuscript.

Funding: This work was financially supported by the Beijing National Natural Science Foundation of China under Grant No. 4222082, the Youth Innovation Promotion Association of CAS under grant No. E1YQXR006, and the National Natural Science Foundation of China under Grant No. 62004216.

Institutional Review Board Statement: Not applicable.

Informed Consent Statement: Not applicable.

Data Availability Statement: The data presented in this study are available on request from the corresponding author.

Acknowledgments: The icac core CMOS program members, the icac pilot line and testlab are acknowledged for their support.

Conflicts of Interest: The authors declare no conflict of interest.

References

1. International Technology Roadmap for Semiconductors. Available online: http://public.itrs.net/ (accessed on 9 February 2022).
2. Vivet, P.; Thuriès, S.; Billoint, O.; Choisnet, S.; Lattard, D.; Beigné, E.; Batude, P. Monolithic 3D: An alternative to advanced CMOS scaling, technology perspectives and associated design methodology challenges. In Proceedings of the 2018 25th IEEE International Conference on Electronics, Circuits and Systems (ICECS), Bordeaux, France, 9–12 December 2018; pp. 157–160.
3. Lu, C.-M.; Deprat, F.; Fenouillet-Beranger, C.; Batude, P.; Garros, X.; Tsiara, A.; Leroux, C.; Gassilloud, R.; Nouguier, D.; Ney, D. Key process steps for high performance and reliable 3D Sequential Integration. In Proceedings of the 2017 Symposium on VLSI Technology, Kyoto, Japan, 5–8 June 2017; pp. T226–T227.
4. Vandooren, A.; Witters, L.; Franco, J.; Mallik, A.; Parvais, B.; Wu, Z.; Walke, A.; Deshpande, V.; Rosseel, E.; Hikavyy, A. Sequential 3D: Key integration challenges and opportunities for advanced semiconductor scaling. In Proceedings of the 2018 International Conference on IC Design & Technology (ICICDT), Otranto, Italy, 4–6 June 2018; pp. 145–148.
5. Fenouillet-Beranger, C.; Brunet, L.; Batude, P.; Brevard, L.; Garros, X.; Casse, M.; Lacord, J.; Sklenard, B.; Acosta-Alba, P.; Kerdiles, S. A Review of Low Temperature Process Modules Leading Up to the First (≤500 °C) Planar FDSOI CMOS Devices for 3-D Sequential Integration. *IEEE Trans. Electron Devices* **2021**, *68*, 3142–3148. [CrossRef]
6. Vandooren, A.; Wu, Z.; Parihar, N.; Franco, J.; Parvais, B.; Matagne, P.; Debruyn, H.; Mannaert, G.; Devriendt, K.; Teugels, L. 3D sequential low temperature top tier devices using dopant activation with excimer laser anneal and strained silicon as performance boosters. In Proceedings of the 2020 IEEE Symposium on VLSI Technology, Honolulu, HI, USA, 16–19 June 2020; pp. 1–2.
7. Micout, J.; Lapras, V.; Batude, P.; Fenouillet-Beranger, C.; Lacord, J.; Sklenard, B.; Mathieu, B.; Rafhay, Q.; Mazzocchi, V.; Colinge, J.-P. High performance low temperature FinFET with DSPER, gate last and Self Aligned Contact for 3D sequential mtegration. In Proceedings of the 2017 IEEE International Electron Devices Meeting (IEDM), San Francisco, CA, USA, 2–6 December 2017; pp. 32.32.31–32.32.34.
8. Yang, C.-C.; Hsieh, T.-Y.; Huang, P.-T.; Chen, K.-N.; Wu, W.-C.; Chen, S.-W.; Chang, C.-H.; Shen, C.-H.; Shieh, J.-M.; Hu, C. Location-controlled-grain technique for monolithic 3D BEOL FinFET circuits. In Proceedings of the 2018 IEEE International Electron Devices Meeting (IEDM), San Francisco, CA, USA, 1–5 December 2018; pp. 11.13.11–11.13.14.
9. Hsieh, P.-Y.; Chang, Y.-J.; Chen, P.-J.; Chen, C.-L.; Yang, C.-C.; Huang, P.-T.; Chen, Y.-J.; Shen, C.-M.; Liu, Y.-W.; Huang, C.-C. Monolithic 3D BEOL FinFET switch arrays using location-controlled-grain technique in voltage regulator with better FOM than 2D regulators. In Proceedings of the 2019 IEEE International Electron Devices Meeting (IEDM), San Francisco, CA, USA, 7–11 December 2019; pp. 3.1.1–3.1.4.
10. Huang, C.-Y.; Dewey, G.; Mannebach, E.; Phan, A.; Morrow, P.; Rachmady, W.; Tung, I.-C.; Thomas, N.; Alaan, U.; Paul, R. 3-D self-aligned stacked NMOS-on-PMOS nanoribbon transistors for continued Moore's law scaling. In Proceedings of the 2020 IEEE International Electron Devices Meeting (IEDM), San Francisco, CA, USA, 12–18 December 2020; pp. 20.26.21–20.26.24.

11. Liu, Y.-W.; Hu, H.-W.; Hsieh, P.-Y.; Chung, H.-T.; Chang, S.-J.; Liu, J.-H.; Huang, P.-T.; Yang, C.-C.; Shen, C.-H.; Shieh, J.-M. Single-Crystal Islands (SCI) for Monolithic 3-D and Back-End-of-Line FinFET Circuits. *IEEE Trans. Electron Devices* **2021**, *68*, 5257–5262. [CrossRef]
12. Deshpande, V.; Djara, V.; O'Connor, E.; Hashemi, P.; Morf, T.; Balakrishnan, K.; Caimi, D.; Sousa, M.; Fompeyrine, J.; Czornomaz, L. Three-dimensional monolithic integration of III–V and Si (Ge) FETs for hybrid CMOS and beyond. *Jpn. J. Appl. Phys.* **2017**, *56*, 04CA05. [CrossRef]
13. Deshpande, V.; Djara, V.; O'Connor, E.; Hashemi, P.; Balakrishnan, K.; Sousa, M.; Caimi, D.; Olziersky, A.; Czornomaz, L.; Fompeyrine, J. Advanced 3D monolithic hybrid CMOS with sub-50 nm gate inverters featuring replacement metal gate (RMG)-InGaAs nFETs on SiGe-OI fin pFETs. In Proceedings of the 2015 IEEE International Electron Devices Meeting (IEDM), Washington, DC, USA, 7–9 December 2015; pp. 8.8.1–8.8.4.
14. Chang, S.-W.; Lu, T.-H.; Yang, C.-Y.; Yeh, C.-J.; Huang, M.-K.; Meng, C.-F.; Chen, P.-J.; Chang, T.-H.; Chang, Y.-S.; Jhu, J.-W. First Demonstration of Heterogeneous IGZO/Si CFET Monolithic 3-D Integration With Dual Work Function Gate for Ultralow-Power SRAM and RF Applications. *IEEE Trans. Electron Devices* **2022**, *69*, 2101–2107. [CrossRef]
15. Chakraborty, W.; Ye, H.; Grisafe, B.; Lightcap, I.; Datta, S. Low Thermal Budget (<250 °C) Dual-Gate Amorphous Indium Tungsten Oxide (IWO) Thin-Film Transistor for Monolithic 3-D Integration. *IEEE Trans. Electron Devices* **2020**, *67*, 5336–5342. [CrossRef]
16. Vandooren, A.; Franco, J.; Parvais, B.; Wu, Z.; Witters, L.; Walke, A.; Li, W.; Peng, L.; Deshpande, V.; Bufler, F.M. 3-D sequential stacked planar devices featuring low-temperature replacement metal gate junctionless top devices with improved reliability. *IEEE Trans. Electron Devices* **2018**, *65*, 5165–5171. [CrossRef]
17. Vandooren, A.; Franco, J.; Wu, Z.; Parvais, B.; Li, W.; Witters, L.; Walke, A.; Peng, L.; Deshpande, V.; Rassoul, N. First Demonstration of 3D stacked Finfets at a 45 nm fin pitch and 110 nm gate pitch technology on 300 mm wafers. In Proceedings of the 2018 IEEE International Electron Devices Meeting (IEDM), San Francisco, CA, USA, 1–5 December 2018; pp. 7.1.1–7.1.4.
18. Lepselter, M.; Sze, S. SB-IGFET: An insulated-gate field-effect transistor using Schottky barrier contacts for source and drain. *Proc. IEEE* **1968**, *56*, 1400–1402. [CrossRef]
19. Tsui, B.-Y.; Lin, C.-P. A novel 25-nm modified Schottky-barrier FinFET with high performance. *IEEE Electron Device Lett.* **2004**, *25*, 430–432. [CrossRef]
20. Kaneko, A.; Yagishita, A.; Yahashi, K.; Kubota, T.; Omura, M.; Matsuo, K.; Mizushima, I.; Okano, K.; Kawasaki, H.; Izumida, T. High-performance FinFET with dopant-segregated Schottky source/drain. In Proceedings of the 2006 International Electron Devices Meeting, San Francisco, CA, USA, 11–13 December 2006; pp. 1–4.
21. Khater, M.H.; Zhang, Z.; Cai, J.; Lavoie, C.; D'Emic, C.; Yang, Q.; Yang, B.; Guillorn, M.; Klaus, D.; Ott, J.A. High-k/Metal-Gate Fully Depleted SOI CMOS with Single-Silicide Schottky Source/Drain with Sub-30-nm Gate Length. *IEEE Electron Device Lett.* **2010**, *31*, 275–277. [CrossRef]
22. Ko, C.; Chen, H.; Wang, T.; Kuan, T.; Hsu, J.; Huang, C.; Ge, C.; Lai, L.; Lee, W. NiSi Schottky barrier process-strained Si (SB-PSS) CMOS technology for high performance applications. In Proceedings of the 2006 Symposium on VLSI Technology, Honolulu, HI, USA, 13–15 June 2006; pp. 80–81.
23. Chin, Y.; Pey, K.; Singh, N.; Lo, G.; Tan, L.; Zhu, G.; Zhou, X.; Wang, X.; Zheng, H. Excimer laser-annealed dopant segregated Schottky (ELA-DSS) Si nanowire gate-all-around (GAA) pFET with near zero effective Schottky barrier height (SBH). In Proceedings of the 2009 IEEE International Electron Devices Meeting (IEDM), San Francisco, CA, USA, 7–9 December 2009; pp. 1–4.
24. Nishisaka, M.; Matsumoto, S.; Asano, T. Schottky source/drain SOI MOSFET with shallow doped extension. *Jpn. J. Appl. Phys.* **2003**, *42*, 2009. [CrossRef]
25. Schroder, D.K. *Semiconductor Material and Device Characterization*; John Wiley & Sons: Hoboken, NJ, USA, 2015.
26. Chen, L.J. *Silicide Technology for Integrated Circuits*; Iet: London, UK, 2004; Volume 5.
27. Qiu, Z.; Zhang, Z.; Ostling, M.; Zhang, S.-L. A comparative study of two different schemes to dopant segregation at NiSi/Si and PtSi/Si interfaces for Schottky barrier height lowering. *IEEE Trans. Electron Devices* **2007**, *55*, 396–403. [CrossRef]
28. Luo, J.; Qiu, Z.-J.; Zhang, Z.; Östling, M.; Zhang, S.-L. Interaction of NiSi with dopants for metallic source/drain applications. *J. Vac. Sci. Technol. B Nanotechnol. Microelectron. Mater. Processing Meas. Phenom.* **2010**, *28*, C1I1–C1I11. [CrossRef]
29. Zhang, Z.; Qiu, Z.; Liu, R.; Ostling, M.; Zhang, S.-L. Schottky-barrier height tuning by means of ion implantation into preformed silicide films followed by drive-in anneal. *IEEE Electron Device Lett.* **2007**, *28*, 565–568. [CrossRef]
30. Lu, J.; Miles, D.; Zhao, J.; Gurba, A.; Xu, Y.; Lin, C.; Hewson, M.; Ruan, J.; Tsung, L.; Kuan, R. A novel nickel SALICIDE process technology for CMOS devices with sub-40 nm physical gate length. In Proceedings of the Digest. International Electron Devices Meeting, San Francisco, CA, USA, 8–11 December 2002; pp. 371–374.
31. Kim, J.; Park, T.; Lee, S.; Lee, S.; Kim, C.; Hyun, S.; Kim, Y.; Hwang, K.; Kim, H. Effect of Millisecond Annealing Temperature of Ni 1-x Pt. x Si Formation on Leakage Current Characteristics of Static Random-Access Memory Cells. *IEEE Trans. Electron Devices* **2018**, *66*, 389–394. [CrossRef]

MDPI AG
Grosspeteranlage 5
4052 Basel
Switzerland
Tel.: +41 61 683 77 34

Nanomaterials Editorial Office
E-mail: nanomaterials@mdpi.com
www.mdpi.com/journal/nanomaterials

www.ingramcontent.com/pod-product-compliance
Lightning Source LLC
Chambersburg PA
CBHW042009260926
51850CB00020B/2943
* 9 7 8 3 7 2 5 8 4 2 8 5 8 *